SELLING IS SIMPLY...
HELPING PEOPLE

SELLING IS SIMPLY...
HELPING PEOPLE

by Daktronics Employee Communications
and Chuck Cecil

Copyright © 2008
Daktronics, Inc.

ALL RIGHTS RESERVED

This work may not be used in any form,
or reproduced by any means,
in whole or in part,
without written permission
from the publisher.

ISBN: 978-1-57579-382-5

Library of Congress Control Number: 2008930731

On the cover: Austin Leite
Cover Photo by Jason Kuhl

Printed in the United States of America

PINE HILL PRESS
4000 West 57th Street
Sioux Falls, South Dakota 57106

Introduction

In 2007, as Daktronics Sales and Service was approaching twenty years in business (1988–2008), Marlo Jones was thinking of a way to communicate the development and growth of the regional office concept and how that concept offers local service and lasting solutions to our customers.

That's how the vision of this book developed.

He thought it would be best to document the history of the offices including the people, places and the experiences of those who founded the offices. The purpose of this book is to acknowledge the efforts of both the staff at corporate and the staff within the offices who had involvement in the success of the sales and service offices.

For those who have participated in this development that began in 1988, this book will be a remarkable trip down memory lane. For those who have more recently joined the Daktronics family, this book introduces you to the Daktronics Sales and Service history and illustrates the rock solid commitment that exists in Daktronics' culture to provide the best customer service in the industry.

Throughout Daktronics history, company leaders have been good teachers through the process of storytelling—but most of those stories have not been written or recorded for future generations to learn from and enjoy. This book is a great beginning.

Acknowledgments

Many people contributed to the creation of this book and played important roles in this storytelling endeavor.

We particularly thank everyone who, by generously sharing their stories and photos and devoting their time to the storytelling process, made this book come to life as an invaluable source of inspiration and corporate history.

We express appreciation to Marlo Jones for the idea and the vision of this book and to Al Kurtenbach and Carla Gatzke for their time and guidance as this project came to fruition.

Thank you to Chet Feil who provided historical documents, photos and detailed stories. He has always been a valued colleague to Marlo Jones and Daktronics.

Keeping the tradition of offering student employees the opportunity to manage pertinent work responsibilities and to experience the pride that is felt when working on a project from the early development stages to the final product, we thank the following employees who conducted numerous interviews with regional employees and then recorded their stories in writing in an effort to inspire, influence and teach: Carl Deardoff, Sarah Even, Lyndi Hawke, John Nelson, Jen Oolman and Caitlin Osborne. We also thank Julie Frank and Amber Nelson who joined the Employee Communications team mid-project and eagerly assisted in administrative roles in the preparation of the manuscript.

Thank you to Jennifer Miller who designed the cover and book graphics.

We extend appreciation to Brian J. Gatzke and the Drake University Business Department for granting us permission to use "The Feasibility of Establishing a Company-Owned Dealership" study.

– Staci Perry, Daktronics Communications Manager
– Chuck Cecil, Author and Freelance Writer

Table Of Contents

1. Finding A Way 1

2. Keeping Up with the Jones 25

3. Getting Started 43

4. Hall Passes and the Secret Service 63

5. The White House Calling 73

6. Pyrotechnics 83

7. Wrestling, Wedding and WordPerfect 89

Going the Extra Mile 113
    Rocky Mountain Region Interviews 113
    California/Hawaii Region Interviews 138
    North Central Region Interviews 145
    South Central Region Interviews 173
    Great Lake Region Interviews 194
    New England Region Interviews 215
    Mid-Atlantic Region Interviews 225
    Southeast Region Interviews 233

Epilogue 247

About the Authors 249

1

FINDING A WAY

The National Coast and Geodetic Survey has fixed the exact center of the United States, if you include Alaska but exclude Hawaii, on a barren stretch of western South Dakota prairie near Castle Rock.

Give or take about 400 miles to the east, Brookings, South Dakota—the home of Daktronics—is close enough to the nation's center even in a game of nationwide horseshoes. This is good if your business includes selling and servicing electronic scoreboards and other dazzling visual displays to the national and international market.

The miles from Brookings to New York or Los Angeles are about equal. It's about that same distance from the Daktronics campus in Brookings to deliver a company product or meet with a potential customer in Miami. So helping customers or selling product in such places as Weslaco, Texas, Rocky Hill, Connecticut, or Chatsworth, California, is a bit of a reach and could be a challenge from the "almost" center of the nation.

But not for Daktronics. The location near the center of the nation is considered a positive. It's a welcome place for the skilled technicians, engineers, and the sales staff representing the Daktronics home office in Brookings. With forty years of experience, the company at the epicenter of the nation has, through trial and sometimes error, crafted a unique national sales and service interlacing that is second to none in the industry.

That network staffed by dedicated employees has helped move Daktronics to eminence as the world's largest manufacturer and supplier of electronic scoreboards, computer programmable displays, large screen video displays, sound systems, and control equipment for audio-visual communication and entertainment systems. It has a sales and service system second to none staffed by the hundreds of dedicated engineers, sales-

This is an aerial view of Daktronics' campus in Brookings, South Dakota, taken during the summer of 2007.

persons, electronic technicians and others in nearly fifty well-supplied sales and service offices throughout the United States just waiting to help. The Daktronics reach also extends beyond the shores and is truly international in scope.

For all of its more than 3,000 employees, helping is

The House voting console in Seattle, Washington.

their mantra. It's the company's maxim, and an important part of its forty-year culture. Customer service and help is just a phone call away and it's always been that way. Even during the early years, in submitting the winning bid for one of the company's first successes—the Utah Legislature's electronic voting system—the promise of after-installation service helped close the deal over a more experienced competitor.

As business expanded incrementally into other areas in those early years of the 1970s and early 1980s, Daktronics began adding products such as scoreboards and time and temperature displays. Daktronics learned the value of more reactive, customer-friendly sales and service procedures to serve the growing list of customers in communities large and small, representing corporations, major league sports venues, high schools, colleges and universities, financial institutions, mom and pop businesses, and transportation terminals and turnpikes. Daktronics products can be found in countless locations, from city parks, church lawns and ice cream shops to center courts and end zones in elementary and high schools around the nation. To sell to and serve these various entities, especially in the early, formative years, Daktronics learned that just getting a foot in the door was the hard part. After that, selling the company product was much easier.

In the very early years, Daktronics' management personnel wore many hats. They were also the sales staff. They worked in concert with independent dealers or resellers. These were small businesses that serviced all types and brands of scoreboards and other electronic signage, and they might also sell a scoreboard or two in the course of doing what they considered their real work of keeping existing installations operating properly. Many independent dealers coordinated their service calls with the opportunity for sales.

This symbiotic connection did not go unnoticed by Daktronics leadership. Servicing equipment could open selling doors. So as Dr. Aelred Kurtenbach and Dr. Duane Sander guided their new company to new heights, they dreamed of one day having what they called, for lack of a better description, "company stores" radiating out from Brookings to blanket the nation. In the evolution leading to that day, about sixty independent dealers assisted in servicing and selling Daktronics products.

But even with this service and sales combination in place, the ultimate goal was a company-owned, company staffed, company coordinated network of sales and service offices. Reaching that goal didn't magically happen. It required considerable thought and planning, accord-

Dr. Aelred Kurtenbach and Dr. Duane Sander with part of their first electronic voting system that was installed in the Utah Legislature.

ing to Kurtenbach. And it took some failures and false starts along the way before just the right mix was devised and a successful recipe was hammered out that met the high standards Daktronics had insisted upon since its first year of operation in 1968. In the words of James Morgan, one of the original employees of Daktronics who became its CEO in 2001, the sales and service network "evolved over time."

Jim Morgan, CEO of Daktronics

Coming up with the right sales and service model was probably much like the quest in the late 1830s to come up with a farm tractor design. No one really knew what a farm tractor should look like. Everyone, however, knew what they should be able to do. Fitting wheels, motors, and the other parts to a frame in the best configuration to do it best job was the challenge. Tractors, in their final basic design, were very familiar in all of their positive and negative aspects to Kurtenbach and Sander, two electrical engineering professors who had the vision of establishing an electronics company that at first was geared to design and manufacture medical devices.

Washington House voting system

Both men, then in their early thirties, had spent hours of their youth operating their fathers' tractors. They knew what tractors could do, and they also knew that tractors sometimes wouldn't start, broke down, and needed services at the most inopportune times in the business of farming.

So when they were cranking up Daktronics in 1968, they remembered, of all things, tractors. They wanted to insure that their new company would back up with service the products they manufactured, just as the tractor dealers in their home towns of Dimock and Howard, South Dakota, had backed up their products for their fathers. So even in the early Daktronics years, a worthy goal was a web of sales and service offices where personnel would not only sell scoreboards, but would also keep them running.

As the company's product line grew and diversified, from electronic voting systems to its innovative Matside® wrestling scoreboards and then the time and temperature displays, and the gigantic marvels in venues such as New York's Times Square and the Texas Longhorns football stadium, so, too did the requirements necessary to respond to product-related emergencies.

Darrell K. Royal – Texas Memorial Stadium at University of Texas.

 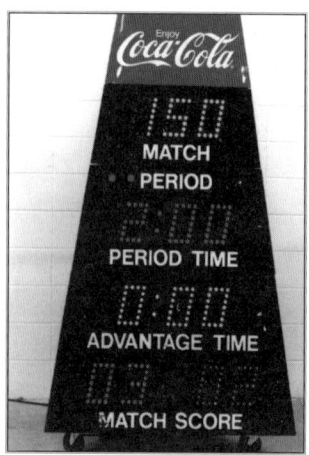

The Coke Display in Times Square.   Matside® Wrestling Scoreboard.

Kurtenbach and Sander had experience with the perplexities of unexpected equipment malfunctions. Both grew up on South Dakota farms that, interestingly, were only about 100 miles apart. They were aware then of the importance of proper and timely equipment maintenance, and they learned the importance to the farming business of rapid and effective customer service.

"Whenever my father bought a tractor in town I was always impressed with the implement dealer's visit to our place, and how he would carefully handle the details of the purchase," Sander recalled. "He always greeted my father by name and spent time visiting with him. It seemed to me that those who were the area's successful business people took a genuine interest in our farm's needs and were curious and inquiring about how we were doing."

These and other experiences helped etch in the minds of both men the importance of customer service. Kurtenbach was first a tractor jockey growing up on the farm near Dimock, South Dakota. He then was an electronics technician servicing sophisticated radar installations in Korea and this country. After Air Force service, while attending the School of Mines and Technology in Rapid City, South Dakota, he worked part-time as a television repairman to augment his monthly GI bill check, and this provided him with another taste of the value of good customer service.

Incandescent football scoreboard for James Madison University.

HD-16 LED video display for the Arizona Diamondbacks.

The co-founders of Daktronics in 1968 zeroed in on three basic business precepts. First, the company needed to be on the cutting edge of technology and be willing to accept the new and abandon the old at the proper times in the company's growth. Today, millions of dollars are invested annually in product development to make what Daktronics manufactures even better and more reliable and to keep pace with a changing world.

Second, Daktronics would be cognizant of the need to operate at the lowest possible cost while providing the very best in product quality. And finally, Daktronics needed to include "customer service" as one of its main premises, an important third leg on the business plan, because no matter the care and engineering invested in the design and building of any product, nothing can be made to operate uninterrupted forever. So the need to always provide excellent, prompt and dependable customer service is embossed and imbedded into the Daktronics culture. The firm's phenomenal success and enviable reputation earned over the past four decades is a testament that none of the three markers set down in 1968 have been abandoned.

Every employee of the company is aware that providing help to the customer is one of his or her primary responsibilities. They welcome the opportunity. "All of the employees at Daktronics just do a terrific job of emphasizing and carrying out customer service responsibilities," Kurtenbach said. "This includes our salespeople, our engineers, our manufacturing people, and especially all of those who comprise our field service staffs."

Kurtenbach believes that the rural, small town background of so many of the company's employees has something to do with its remarkable record of sales and customer service. He likens it to coolness under fire. "From my own experience on the farm I know that farm kids are exposed to the daily uncertainties and unexpected and untimely interruptions of the farming cycle," he said. "Farm kids learn in real world situations how to deal with emergencies and uncertainties, from unexpected weather conditions to unforeseen and often rather abrupt equipment breakdowns. And in rural America there are always examples of farm families lending a hand to help a distressed neighbor in need," he added.

"When something with a Daktronics logo on it has a problem, because of their early experience with similar situations on the farm, they are able to remain cool, calm, and collected and to think rationally about what needs to be done, and to not become flustered when working under high pressure situations," he said. Also, just as farm kids grew up helping friends and neighbors, he believes they have a built-in, natural desire to help others in challenging situations.

Kurtenbach also believes that those raised in a rural setting make excellent salespeople. "That's because we are more open and we are more likely to be helpful, and these are two important characteristics that lend themselves to good salesmanship. I have always told our potential salespeople that selling is really just trying to help people, and Daktronics management is proud that its employees are always very helpful to others."

Daktronics CEO James Morgan, who grew up on a farm near Luverne, Minnesota, and started with the company as an electrical engineering student at South Dakota State University in the first years of the company, agrees. "Customer service is the area in which we really do differentiate ourselves from others—not just our competitors—but others who serve the same market we serve," Morgan said.

During the early 1970s, the company, almost in a serendipitous moment, found its niche not in medical instrumentation, but in sports scoreboards, and then developed its various in-house management systems, realizing that external sales and service capabilities deserved a high priority. Kurtenbach and others in the company, then about fifty employees strong, did what they could to introduce and sell their product. To help, they also utilized independent dealers to sell and service, but this arrangement was often akin to hit and miss.

As part of that effort, whenever a dealer or a customer contacted Daktronics with a product problem or a question, Ed Weninger set aside his work to take the call. He had joined the company in 1970 fresh out of the North Dakota College of Science and Technology with a degree in Industrial Electrical Technology. He was a talented and skilled electronics technician for the company, working in product development, and either knew the answer to the question, or knew where to find it. Those early contacts and calls to Weninger's desk for help were the genesis of the company's eventual customer service effort, now an important rung in the Daktronics ladder.

Aware of Weninger's customer service experience as well as his technical expertise and talent with salesmanship, Kurtenbach decided to involve him when a small-scale company sales and service venture was tried in a ten-state Upper Midwestern region. Weninger had often worked in the five-state area of South Dakota, Minnesota, Iowa, Nebraska, and Kansas with Kurtenbach, Larry Fjeldos, and a fellow North Dakota College of Science alumnus Pat Schwan.

Ed Weninger in his Dekalb, Illinois office (Spring 1979).

They contacted schools for possible scoreboard sales, and called on banks and other financial institutions to sell the Daktronics time and temperature signs. Weninger remembers

Ed Weninger on what he calls the "never-ending" roads of Iowa.

that an average week on the road for him started on Sunday and ended the next Friday evening or Saturday morning, driving town-to-town making the cold calls, telling the potential clients about Daktronics and leaving them with a not-too-fancy brochure. "My future wife Deanne and I had what I called our 'weekend courtships,'" Weninger said. The roads seemed never-ending and he still remembers the day he drove into Carroll, Iowa, because with that stop, he had visited every western Iowa town on the map.

One of those early fellow travelers, Pat Schwan, was later asked to establish a Daktronics presence in Illinois. So Schwan worked out of Bloomington, Illinois, before leaving the company in 1977. To fill the void left by his departure, Weninger agreed to move to Illinois. With his wife Deanne along on a scouting trip, they decided to rent an apartment in DeKalb. Weninger still remembers the discussion he had with Kurtenbach about the move. A competitor, American Sign and Indicator Company, headquartered in Spokane, Washington, had a regional office in Illinois. "So we got out an old atlas," Weninger said. "Al was always looking for an edge, and we talked about our location in South Dakota being an advantage to us in Illinois over a company on the West Coast like American Sign and Indicator Company."

Working conditions were tight in an apartment.

In 1977, Weninger and Deanne drove the company-provided five-year-old Plymouth loaded with all of their worldly belongings and Weninger's toolbox, and headed east. "I was excited about the move and saw it as a challenge," Weninger remembered. He said he suspected his wife,

Greg Thorness taking a call from his apartment office.

a small town girl, wasn't all that thrilled about it, but she never complained until once on a trip passing through the hustle and bustle of Chicago. The couple settled into their little apartment in DeKalb. Their one-car garage became the Daktronics storeroom and workshop. On cold days, Weninger often brought his work to their two-bedroom apartment and spread parts and tools out on the bed. Their telephone had one of the first-ever answering systems attached, and Deanne would often take the business calls.

On the road, before the cell phone was a common utility, Weninger's contact with the outside world was the pay phone. He soon learned where most of the booths in Illinois were situated. If he needed a ladder or a larger piece of equipment, he'd rent or borrow from someone. Weninger made cold calls in Illinois, Wisconsin, and Indiana.

Often Deanne would travel with him. She remembers a trip they made to Madison, Wisconsin. She was pregnant at the time. The day after they got back to their DeKalb apartment, their first daughter Lanette was born.

Among Weninger's memories is being approached by Union stewards dressed to the nines, smoking big cigars and arriving unexpectedly at wherever he might be installing a Daktronics sign. They drove up in fancy cars, approached him and usually asked personal questions, wanting to know what his salary was for installing time and temperature signs. The farm kid from near the late bandleader Lawrence Welk's hometown of Strasburg, North Dakota, was bold in telling those bejeweled Union representatives that his salary was none of their business. With that reception, they lost interest and drove away, Weninger remembered.

Later, the Daktronics office sent technicians to help Weninger, young men like Gary McDonald, Steve Lawler, and Greg Thorness, to name a few. In 1981, Weninger asked to be transferred back to Brookings. Jeff Robbins was sent to Illinois to pick up where Weninger left off. Weninger settled in at the Brookings office where one of his first assignments was to formalize the customer service system he knew so well and had actually started before his Illinois assignment. At that time, Daktronics had about 100 employees. Weninger is currently the company's Manager of Customer Service.

The lessons learned by Weninger and others on those early sales forays to augment the work of what resellers and independent dealers were doing, while admirable, left something to be desired. The system lacked continuity, was inefficient, attempted to cover too much territory, and was organizationally awkward. Kurtenbach and company officials searched for another way to sys-

From left going clockwise around the table: Jeff and Gudrey Robbins, Ed Weninger, Deanne Weninger and Gary McDonald.

tematically sell products nationally in the "company store" mode. In the summer of 1979, Kurtenbach had decided to call for help from his younger brother, Frank, intending for him to assume the responsibilities of National Sales Manager for Standard Products.

Frank Kurtenbach graduated from South Dakota State University in 1961 and earned his master's degree there in 1969. While an undergraduate student, Kurtenbach was a champion heavyweight wrestler, and after graduation his interest in sports took him to Muscatine (Iowa) High School where he was a coach and taught biology for two years. He then moved to Keokuk, Iowa, for three years with similar responsibilities on the field, the mat, and in the classroom. After three years at Keokuk, he returned to his alma mater for his master's degree in physical education. A temporary vacancy came up and he was asked to also coach the university's wrestling team for one season. His master's thesis was on various aspects of high school wrestling rules. His study caught the attention of officials of the state's wrestling association, and this in turn garnered him an appointment on the state wrestling association's rules committee.

This later led to his appointment to the national high school wrestling association's rules committee. After earning his second degree, Kurtenbach was hired as the head wrestling coach and biology teacher at Lincoln High School in South Dakota's largest city, Sioux Falls.

During his tenure in Sioux Falls, having apparently inherited some of his older brother's talent for entrepreneurship, he established a company called Great Plains Mat Refinishing. He spent six summers traveling to schools in the area refurbishing wrestling mats, and he also dabbled in the Sioux Falls real estate market when not teaching and coaching. His older brother at Daktronics was impressed with his business acumen, his service on the national rules committee, and his large circle of friends and acquaintances in the athletic world. All these, coincidentally, would be helpful to Daktronics, especially in scoreboard sales initially, and in selling the company's other product lines eventually.

It proved to be an insightful hire for Daktronics. The younger Kurtenbach immediately lent his talents and his knowledge of sports and high school athletics governance at about the same time as the firm's new and novel Matside® wrestling scoreboard was making inroads nationally. It also came along as the firm's line of scoreboards for other sports was growing and gaining acceptance. Kurtenbach's experiences at the high school and collegiate sports level were invaluable. He could "talk the talk," and he understood the lay of the land at both the high school and collegiate sports level. His work would help move the evolving Daktronics sales and service compass needle in the right direction.

Frank Kurtenbach talking to employees.

But Frank Kurtenbach admits it took time for him to become acclimated to the business world after years in public education. And electronics wasn't exactly his cup of tea, either, so he had to also brush up on the technical side. He remembers his first day on his new job in August of 1979. He'd been at his desk only a couple of hours when someone from Colorado called his secretary, Jenny Erickson, with a question. The caller wanted to know why did the Daktronics ST-210 timer cost more than the ST-213. The National Sales Manager didn't know at that time that the last two digits in the numbers indicated the size in inches of the timer numbers. The 213 had a larger face than the 210, but was less expensive. Kurtenbach proved to be quick on his feet for a former heavyweight wrestler. He told the caller that he was away from his desk and would call back in a few minutes. He set out to find someone at Daktronics who knew the answer to the caller's question, and then made the return call.

As time went by he learned the language and all about Daktronics products. With his brother, Aelred and other Daktronics officials involved, a series of reviews started to determine what was needed to continue the evolution of the company's efforts at establishing satellite sales and service offices. It became obvious, among other things, that mixing apples, the sports scoreboard side of the sales and service effort, with oranges, the financial entities that were the primary market for the time and temperature displays, had inherent problems.

Daktronics leadership determined that they were dealing with different entities as if they were the same. A chasm existed between the market for the sports side of the business and the market for the business side at which the time and temperature displays were oriented. A lesson learned in that first foray was that a good share of the scoreboard market was in the public school sector, both K-12 and the collegiate levels. These public entities operated in a different pattern than did the private business

Spring Valley Bank

George F. Emma Jr. Memorial Field

sector. The schools had to adhere to different rules and regulations for purchases compared to the private business sector. Different lead times of the two sectors were involved, and there were unique sets of urgencies as well. "We might submit a proposal to a public school and it would be filed away and we might not hear from them again for a year," Al Kurtenbach recalled. "When we submitted a proposal to a business, they might tell us to give them a day or two or a week, and they'd get back to us."

Because of these challenges, a change was needed. It was back to the drawing board. Kurtenbach asked his son-in-law, Brian Gatzke, to help. Gatzke was a former sales intern hired by Gary Gramm who was later promoted to head up the company's efforts into the high school and parks and recreation market. At the time of the study Gatzke was enrolled in a master's degree program at Drake University and needed a project. Kurte-

```
                    RISKS OF THE INDUSTRY
Government Intervention
        1.  tax structure changes
        2.  outdoor sign codes
        3.  license and permits
        4.  product safety codes
        5.  distribution of funds
Competition
        1.  new products
        2.  price fluctuations
        3.  warranty fluctuations
        4.  technological changes
        5.  hiring of trained personnel
        6.  changes in methods of selling
Economy
        1.  fluctuations in key industrial components used in
        product, (availability, price, etc.)
        2.  world-wide political policies (wars, embargoes)
        3.  regional catastrophes
        4.  major employers leaving areas or closing operations
Demographics
        1.  population growth and expansions
Private Donations
        1.  changes in beverage companies policies
        2.  reduction in funds due to economy changes
        3.  public facilities not allowing advertising
```

"The Risks of the Industry" section from The Feasibility of Establishing a Company-Owned Dealership paper by Brian J. Gatzke and Drake University Business Department.

nbach asked him to take a new look into the feasibility of establishing a network of Daktronics-owned offices in the Upper Midwest. Gatzke's study identified pitfalls and some of the opportunities. His investigation was the necessary groundwork for the company sales and service system now in place.

Historic Scoreboard Sales and Service logo.

From Gatzke's study and after hours of discussion among Daktronics leaders, it was determined that a combination on-site salesperson and technician met the company's need. This created the spark of an idea for a successful small scoreboard sales and service organization. Servicing equipment, regardless of make or model, servicing *all* equipment made by any competitor, would open

Gary Gramm

the doors for future selling opportunities. It was a joined-at-the-hip way for Daktronics personnel in the field to become acquainted with potential customers and provide them with service they would appreciate and remember. By following this "service first" model, it was predicted Daktronics personnel in the field would generate a cash flow to provide for most of the office's expenses in its formative years.

A faint hint of this core idea had surfaced for Daktronics years before when Frank and Aelred had enlisted the help of Gary Gramm, who was then working at nearby South Dakota State University attempting to establish cooperative education or paid internship arrangements for students that related to their major field of study. To promote its Matside® wrestling scoreboard to the high school, college, and university wrestling world, Daktronics offered to provide the scoreboards at high profile wrestling meets, big tournaments, and national wrestling venues. At first, Daktronics officials hit the road to showcase the amazing new scoreboards. They saw to the scoreboard's proper installation and servicing if necessary. Even company co-founder Kurtenbach volunteered for the circuit. "It was fun for a year or two, but over time it became very demanding and also quite expensive to have company officials on the road for days and possibly weeks at a time," he said. A student intern program was needed and could relieve the need for management to hit the tournament road.

Early demonstration trailer.

With Gramm's help and expertise, Daktronics mapped out a company sales and service internship program that has stood the test of time and has proven very successful. Many present Daktronics regional office managers recall those long trips to wrestling matches, and later, showing off other Daktronics scoreboards to potential customers. It became a part of their customer relations training. Coincidentally, the sales and service internship model fit somewhat into what Kurtenbach and Sander had envisioned in their first business plan. The students at South Dakota State University in Brookings, where both were teaching, were not only bright and talented and in need of part-time jobs, but were potential assets—future employees—for the company.

The intern model that Gramm designed would initially involve students in the university's engineering programs, but collegians in other fields of study, such as computer science, business, and other pursuits could also benefit from the program as structured by Gramm. And of course, Daktronics would be better because of it, too. These young men and women were a dependable and creative, part-time labor force from which might emerge future employees with special talents, skills, and exuberance. The intern program was similar to team try-outs. Students learned if the world of manufacturing was for them, and Daktronics could identify students it believed would be excellent additions of its full-time team.

The first sales and service intern Daktronics hired was Jeff Robbins, who would later succeed Ed Weninger at his DeKalb, Illinois, office. Robbins is presently the company's Purchasing Manager. He grew up in the Brookings area and graduated from Brookings High School. He was a student at South Dakota State University majoring in business when he was hired in January of 1980 as the first student sales and service intern to help in the introduction of the Matside®. Before he was sent out on the wrestling circuit to show and promote the product, he spent several weeks of orientation in the plant helping to build the scoreboard so he would know the product inside and out.

He remembers that his experience on the Matside® hustings was both enjoyable and educational. After a semester showing off the scoreboard, he returned to the university and graduated in 1980. He has now worked for Daktronics for 27 years. To this day he remembers the advice he received from Al Kurtenbach early in his career. "He called me into his office late one afternoon and we visited. During our conversation he told me to be sure to treat everyone who contacted the purchasing office or who visited Daktronics the same as I would have like to have been treated when I was in DeKalb as a salesperson," Robbins said. "It was good advice, and I've followed his suggestion since then."

As the program grew, other interns, including Jim Vasgaard and Brian Gatzke early on, were brought on board. After training in the intricacies of the Matside®, interns drove a Daktronics vehicle with the scoreboard tied down securely in a trailer behind it to the larger wrestling events around the country.

The interns were expected to set up the scoreboards, test them for proper operation, reconfigure their locations as the tournament eliminations progressed to the single-mat finals, and be there should something in the scoreboard system fail, although because of good engineering, this seldom occurred.

During down times from mat action, and at other propitious moments during the event, the

Jeff Robbins

intern would set up a table in the hallway or some other convenient place at the venue, distribute scoreboard literature and answer questions about the Matside® with the hopes of stimulating a sale. "A card table, simple handouts, and an intern's salesmanship may sound rather simple now, but at the time we were pleased to learn that this was a powerful sales tool for us, and we considered it to be a very successful way in which to market that particular item," Kurtenbach remembered. As it turned out, since the intern set up and serviced the scoreboards, then attempted to stimulate sales, it was an early form of the "service first" concept.

Gramm, who was invaluable as the intern program was established, grew up in the tiny South Dakota town of Hosmer, was hired full time in 1988 by Daktronics as manager of the High School and Parks and Recreation department and helped in staffing regional sales and service offices, where the one-two pattern of servicing and later selling was vigorously pursued in the early years of establishing these "company stores." The original premise of service to enhance sales was gradually abandoned as more and more Daktronics displays were sold. Today, the old "service first" concept has been replaced with the new mantra of "sales *and* service first." Continuing the "service first" concept, especially for obsolete scoreboards, would in effect have limited Daktronics' opportunities to sell new, modern scoreboards.

As the evolution for the beginnings of the present-day Daktronics sales and service network matured, there was a need to be cost effective and efficient in the outlying operations. One of the least cost effective parts of any business covering a wide area is the cost of "windshield time," driving to and from destinations. Having people at each "store" reduced this "windshield" cost and left more time for selling, installing and services.

A goal Kurtenbach, Gramm and others decided upon was to establish one new Scoreboard Sales and Service offices (later changed to Daktronics Sales and Service offices) every quarter. It was determined that the first employee in each office would be an electronics service technician who would also sell. As mentioned, all of the sales and service offices were in the early years branded as service-first entities working on any scoreboard no matter the make or model. But of necessity, this has changed to place a bigger emphasis on the sale and service of Daktronics products only. However, in keeping with the company culture of helping others in need, there are times out in the field when a Daktronics technician will lend a helping hand to a client or potential client that has an emergency situation with a competitor's product.

Kurtenbach remembered that his study team also developed a simple plan for entering the market in each new sales and service area. "We decided we would approach the schools and others in what we thought would be a refreshing way," Kurtenbach said. "We would begin this introductory phase by having the technician live in the area and call on the school's officials periodically to explain that we were in the business of servicing all scoreboards, no matter the kind or brand." The technician would take an interest in the school and the community and become involved as an active, interested citizen. This was not only good business, but it was what Daktronics expected of all employees for their own well-being, happiness, and personal growth.

As the relationship between the school and the Daktronics sales and service representative progressed, school officials and other entities, as

Heinz Field is one of Daktronics' "super displays."

the opportunities arose, would be made aware that Daktronics not only serviced scoreboards but also sold Daktronics scoreboards that were superior to the product currently installed and being serviced by the Daktronics technician.

It had often been observed by Kurtenbach and others that with most competing companies, while the sales and service offices were comprised of both salespeople and a cadre of service people, the two groups did not necessarily work as a team. One hand usually didn't know what the other was doing. Often, in fact, the service people worked out of their homes and had different supervisors from the sales side. There was minimal, if any, communication between the two groups at work for the same company. This divisiveness was not a recipe for success. "I wanted to make sure this kind of a situation didn't come to pass with Daktronics," Kurtenbach said.

At Daktronics, the continuum from service to sales and sales to service was considered a circle of opportunity, with each benefiting and complimenting the other. Today each office may have specialists in one type of display or other, but the hope is for a homogenous mix that will pull together when necessary, such as the times when Daktronics is installing one of its super displays that are becoming more and more popular.

So the electronics technician in the new Daktronics Sales and Service offices, who might be an employee who started out as a student intern in the footsteps of pioneer Jeff Robbins, would be the nucleus of the company's initial footprint in each new geographic area. The technician would be the company's VIP, knowledgeable about how all of the various scoreboard and signage products operated and what was necessary to install and repair them. The electronics technician was the one out ahead walking the company's point. On site, the technician was Daktronics and all that it stood for.

Kurtenbach said a positive supporting factor in this approach was Daktronics' growing reputation as a company sincere in its intentions to provide the very best customer service. "It was also known that Daktronics employees were rather hard-working people," Kurtenbach added.

Kurtenbach and other Daktronics officials had put great stock in their plans for a series of company-owned sales and service offices across the nation. The opportunity to launch this new plan, crafted from the lessons learned in the previous sales and service forays, came to Daktronics quite unexpectedly.

2
Keeping Up with the Jones

Late in the afternoon on a hot June day in 1988, an unassuming Seattle, Washington, church deacon and part-time missionary to Africa waited in the Daktronics office lobby for CEO Al Kurtenbach to leave work for the day. The man had the key to the successful launch of what are now nearly fifty Daktronics "company stores" throughout the nation.

Chet Feil, highly respected and well-known independent Seattle scoreboard sales and service businessman had found a new calling—church work in Africa.

He and his wife, Ruth, had spent months in Africa doing God's work. They enjoyed what they were doing but were concerned that their many loyal scoreboard customers in the Seattle area were being short-changed because of the time serving in Africa. Feil also felt it was time to retire and enjoy life and the new avocation he and Ruth had found helping the less fortunate on another continent.

Chet and Ruth Feil in Seattle, Washington.

One of his clients in the Pacific Northwest had been Daktronics, a company that appreciated his knowledge and expertise in repairing and occasionally selling its products. Likewise, Feil had been impressed with the Daktronics personnel with whom he had been in contact. He found the Daktronics products to be among the best he had serviced and sold.

Feil also enjoyed his journeys to the annual Daktronics independent dealers meetings in South Dakota, and especially the company gatherings at the Al and Irene Kurtenbach farm.

Late on a June day in 1988, Feil waited in the lobby of the Daktronics office for Kurtenbach to lock up for the day. When Kurtenbach emerged from his office, Feil rose from his chair to greet Kurtenbach and at that time told him he would be interested in selling his Seattle business to Daktronics.

Feil was tiring of the routine, his wife desired to pursue their work as part-time missionaries in Africa on a more regular basis. Feil told Kurtenbach he was concerned that the time he was spending on his beloved missionary work was taking time away from his Scoreboard Parts and Service business. Kurtenbach expressed an interest and inquired about Feil's selling price. It was $50,000. "I thought it was a reasonable asking," Kurtenbach later recalled, "but I wanted to structure the purchase properly."

Chet showing a Daktronics scoreboard at a trade show.

A summer picnic in Frank Kurtenbach's backyard before summer training at corporate.

At that juncture in the company's history, Daktronics employed about 350 people and had annual sales of nearly $20 million. It's first year of full operation, 1969, had annual sales of $2,352.13.

Kurtenbach had been searching for an approach that might lead to the establishment of a national scoreboard sales and service network—the company stores he had often dreamed about—and the meeting that June

DAKTRONICS INC.

This certifies that

Chester Feil

has met the required standards for completion of the

DAKTRONICS SCOREBOARD SERVICE SEMINAR

as set forth by Daktronics, Inc.,
this   6th   day of   August   , 19 81 .

EDWARD A. WENINGER
CUSTOMER SERVICE - MANAGER

The certificate for Chet Feil completing Daktronics Scoreboard Service Seminar.

day seemed to him to hold promise. So as it turned out, both Daktronics and Chet Feil were able to make some propitious moves in 1988, one to organize the beginnings of a very successful sales and service network and the other to allow time for Feil's calling in Africa.

The Seattle service van (1989).

The first move for Daktronics to launch this national venture was to buy Feil's Scoreboard Parts and Service business in Seattle, Washington. The second move, more necessary than historic, started at Chet Feil's jam-packed garage on southwest Twenty-fourth Street. The moving job fell upon the shoulders of the new Daktronics sales and service representative, young Marlo Jones from the small town of Pollock, South Dakota. He was a graduate in Electronics Engineering Technology from South Dakota State University. He loaded up all of the tools, a multitude of spare parts from just about every make and model of scoreboard ever invented, and a plethora of equipment from step ladders, ropes, and cables to lag bolts and boxes filled with switches and connectors. Jones, whose hobby is the study of the history of WW II and Korea (his father, a Marine in Korea, earned a Purple Heart in the Nevada Battles), particularly remembers the heavy boxes and cans that were full of military-style connectors of various sizes that were used for the infamous Army "Walkie-Talkie" radios from his father's era, and the variety of parts for electro-mechanical scoreboards. Packed in the truck Jones had rented, it all comprised the inventory of Feil's Scoreboard Parts and Service Company. It was the beginnings of Daktronics' very first bona fide Scoreboard Parts and Service office at 309 South Cloverdale Street, suite C-2, Seattle.

Marlo Jones at the front door of the Seattle Scoreboard Sales and Service office in 1988.

Jones got everything organized and ready for business and on December 1, 1988, without trumpets or drum rolls, Daktronics quietly launched its first satellite Scoreboard Parts and Service office. The office would operate under the original name "Scoreboard Parts and Service" until the summer of 1990 when the decision was made to drop "parts" and add "sales." By either name, the office was to become the bellwether for what eventually would be forty-nine similar offices strategically placed across the nation. That first opening was inauspicious to say the least. It was just 900 square feet, including a small warehouse for Jones' tools and parts. It rented out for slightly more than fifty cents a square foot. While the inventory practically filled the space to the brim, the inventory was just a small part of the package Daktronics purchased in October of 1988 for $50,000.

Equipment List	
Description	QTY
Naden BB scoreboard w/control	1
Fairplay BB scoreboard w/control	1
Daktronics BB-1013 scoreboard w/control	1
Nissen BB scoreboard w/control	1
Datec shot clocks (set) w/control + cables	1
Daktronics ST 213S (set) w/control + cables	1
Daktronics ST 213 controller	1
Step ladders 3'	2
32' ladder (extension)	1
3/8" drill reversible	1
Block + tackle + 130' rope	1
Knock out set	1
Assorted tools and sockets	
Assort lift chains (small)	1
Extention cord 100'	1
Nissen test digits	2
Toolbox full of misc switches + relays	1
Nissen testscore paks 4256	1
Scoretime test sure paks 3004	1
Asst. test cables	N/A
MD Brown test set	1
Scoremaster test digits w/extra lampbank	2

Daktronics Inc. | PROJECT | Nº
TITLE Schedule 2.2 | DWN: | DATE:
ED Nº | REVISED: | DES: | CHK: | APP:
DF-1177 | PAGE | FIG

The 1988 partial Scoreboard Parts and Service inventory list.

The $50,000 had also bought the good name, reputation, and the goodwill of Chet Feil and his business. It also included his customer and prospect lists and perhaps the most important of all, Feil himself. He had agreed to delay his church work and remain on the job helping and advising Jones for a year, assisting the new kid on the block as he began building what would become a true Daktronics success story and a growing scoreboard sales and service office in the Pacific Northwest.

Inventory of the Seattle Sales and Service office (1988).

Moving wasn't exactly unfamiliar to Jones, the man selected by Daktronics to pioneer the company's new venture—or adventure—leading to a scoreboard sales and service network. Daktronics CEO, Al Kurtenbach, had taken notice of Jones, first as an intern and then as he settled into his new full-time job with the company after earning his degree from South Dakota State University in 1986. As an eager student intern for the company in 1985, Jones worked under the tutelage of Ed Weninger, who had been in the trenches in an earlier effort at establishing satellite offices, and others with similar backgrounds. He first gained valuable experience in the customer service department repairing modules and computer equipment and making service calls.

Jones was a good technician and had an adventuresome spirit, seeking out and welcoming challenges both at work and at his hobbies of canoeing, hiking, camping and flying airplanes, and Kurtenbach liked that. Jones remembers traveling to Pierre, South Dakota, with others to service the Daktronics electronic voting system there, and moving on to other sites for service calls in several Midwestern states, such as Illinois, Wisconsin, Michigan, and Ohio. He says he gained valuable experience from these forays to other states, since Customer Service at that time also handled some of the in-house technical support including maintenance of the telephone system. "This gave me many opportunities to interface with the sales and support staff in other departments and it was also helpful to me in learning trouble-shooting skills for products of other than our own design," he said.

After graduation, Jones immediately became a member of the customer service staff as an installation supervisor. That meant moving around from one job site to another, gaining experience with the larger installations such as the one at the Clemson University football stadium, the University of Nebraska basketball project, and the Calgary Olympics speed skating project. "When we traveled for installations we would also make other stops for service calls or to meet and support dealers with training or technical help," Jones recalled. Barely out of college, he was thrilled to learn in 1987 that the company was sending him to supervise the installation and support of displays at the All-African Games in Nairobi, Kenya. In 1988 he worked with Al Kurtenbach at the Calgary Winter Olympics and he followed that up later with an exciting trip to Saudi Arabia for more installations.

Back in Brookings and by then supervising the technical area of customer service, he was also afforded the opportunity to work more closely with the knowledgeable and experienced Ed Weninger. "I was also fortu-

Marlo Jones in the van bay of the Seattle office (1988).

Marlo Jones in the Seattle Daktronics Sales and Service office (1988).

nate to be asked to take a supervisory position in customer service which allowed me to participate in the 1988 summer High School Parks and Recreation (HSPR) seminar in Brookings," he said. He had attended previous HSPR dealer seminars as either an intern or trainee and therefore had the opportunity to become acquainted with many of the independent dealers. Additionally, he had occasion to talk with independent dealers as part of his work on the Daktronics help desk phones. As a result of all of this, he became acquainted with Chet Feil. The two hit it off from the beginning. Both were skilled at what they did, and although there was an age difference and they hailed from different parts of the country, they both had a rural orientation and were just country boys at heart. The idea of living in Seattle appealed to Jones, the farm boy who by this time had traveled much of the county and some of the world.

Feil's journey to become a successful, respected independent businessman had been an interesting one. He started out in the 1950s as a machinist for the Gerlinger Carrier Company in Dallas, Oregon, before joining the Boeing Company in Seattle in 1958, also as a craftsman and machinist. While there, he took an evening course and with this training became a Certified Electrician for the Boeing Aircraft Company. As a state licensed electrician, he qualified for a job at the Highline School District in Seattle, where he worked for the next fourteen years. Because of his exceptional talents as an electrician, the Highland School District gave him responsibility over all of the district's hundreds of clocks, which became a challenging job during Daylight Savings time changes, Feil remembered.

He was also assigned the task of maintaining the school's sports scoreboards. He particularly enjoyed the challenge of troubleshooting this type of complicated equipment and he became very good at it. He then accepted a similar position at the nearby Mercer Island School District.

While with the Mercer District, because of his experience as a machinist, an electrician, and one knowledgeable with the complications of Highline school's time systems, and then with the on-the-job expertise gained from maintaining the school scoreboards, he decided he would enjoy and could be successful repairing scoreboards for schools and other educational entities throughout the Seattle area. His plan was to continue his regular job at Mercer, and do the scoreboard repair work on weekends or after work on weekdays.

With a supply of tools and scoreboard parts purchased for $200 from Athletic Supply of Seattle, he

Chet Feil at a trade show.

soon gained a local reputation as an expert in scoreboard repair, able to fix the problems that came up with any make and model, from Nissen and Scoremaster, to Fair-Play, Naden, Nevco, and others. Business was good. When Feil retired from the school district, it became even better because of the additional time he could spend not only repairing, but also selling football and basketball scoreboards.

In addition to the market he'd established for scoreboards in area grade and high schools, Feil branched out and sold to colleges and universities, youth clubs and businesses. Interestingly, he did not advertise his business, depending instead on word of mouth, the ultimate in advertising. Word of his ability and service made the rounds in the Seattle area.

The echoes of Feil and his good work even found their way to Florida, where Frank Kurtenbach of Daktronics was attending an industry meeting. National Sales Manager Kurtenbach heard about Feil at that meeting, and since Daktronics was always on the lookout for good, reliable independent dealers, Kurtenbach was determined to get to know this well-known salesman-technician, and convince him to become an independent dealer and service technician for Daktronics products. Kurtenbach finally found Feil working in his home workshop with scoreboards scattered about and flashing brightly or in various stages of repair and disrepair. During their conversation, Feil inquired if Daktronics could provide him with an ETL-approved (Electronic Testing Laboratory) scoreboard that was required by the State of Washington for a possible scoreboard sale Feil was negotiating. "Frank said he could do that," Feil recalled, "so I took on his scoreboard line to sell and repair." Feil also had the opportunity to work with Daktronics' High School, Parks and Recreation Director Brian Gatzke on a voting system repair assignment in the State Capitol Building in Olympia, Washington.

Through these contacts and others such as Ed Weninger, Marlo Jones, and others, Feil came to be a part of the Daktronics family. He was impressed with Daktronics products and the company's helpful, friendly, down-to-earth personnel.

That June of 1988, Feil was invited back to the Daktronics for a seminar on new products, which also included an invitation to the annual company picnic then held at the Al and Irene Kurtenbach farm near Brookings. "The children even got to try to milk a cow out there, and they played games and we also all had good food," Feil fondly recalled.

It was on this trip back to South Dakota that Feil met with Al Kurtenbach and offered to sell his Scoreboard Parts and Service Company to Daktronics.

The two later agreed on the price provided that it would include a 12-month consulting agreement, a non-compete clause, and purchase of all

Paul Wildeman riding a "bucking barrel" (on the Kurtenbach farm) as Frank Kurtenbach (center right) provides the power.

inventory, the business name, Feil's customer list, and a continuation of the Scoreboard Parts and Service telephone number.

As Kurtenbach drove home that evening after the brief visit with Feil, the name of a young man at the plant came to mind. "I was aware of Marlo Jones because he was working for us in our Customer Service Department and doing a very good job," Kurtenbach remembered. He said he thought Jones was very promotable. "He was an ideal employee and I didn't want him to feel he had reached his potential at Daktronics and would have to leave us and go elsewhere to be challenged."

At the time of the purchase from Feil, there was talk of calling the new satellite the Daktronics Sales and Service office, but because there still remained in that area and across the country independent dealers and sign and repair technicians with whom Daktronics worked, that did not take place. "Some of the companies were a bit antsy about us going out and competing against them," Kurtenbach said, "so we decided it would be better not to include Daktronics in the name and just go by Scoreboard Sales and Service." Over time, this concern faded and the Daktronics name was added.

Kurtenbach and Jones talked over the possibility of Jones moving to Seattle and setting up the new office. Jones flew out to Seattle to get a feel for the area. He saw the potential that existed there. He loved the weather and the lay of the land. "I'll take it," Jones enthusiastically told Kurtenbach. "I still remember my days of driving a van to Seattle with some tools, a few of my personal belongings, and a plan," remembered Jones. He made a quick stop at his parents' home in Pollock to pick up some tools and other items, then drove straight through to Billings, Montana, that first day, making plans in his head as he drove west. "My plan then was not re-

Marlo and his van in Brookings, South Dakota, before leaving to start the Seattle office.

gionalization, but rather to make my work and my office successful. I think that's why Daktronics has done so well—we all focus on doing our best, together."

As Jones' wife Susan was settling into their new apartment, her husband was organizing the Seattle operation, sometimes creating mailings using the then brand new IBM Model 30 personal computer. He, Kurtenbach and Feil were also exchanging ideas and views on upcoming projects in the Pacific Northwest that might be fruitful for Daktronics. "Although the High School and Parks and Recreation (HSPR) market would be our main focus," Jones remembered, "it was clear from the start that the presence of Daktronics in that area would help us in other markets as well." In the years that followed, Jones would discover just how much this Daktronics presence would help with larger projects in all the markets.

Among the upcoming large market projects Jones and Kurtenbach had on their radar screens were the displays that would be needed for the 1990 Goodwill Games aquatics facilities, the Spokane Coliseum hockey scoreboard, several possibilities with Coca Cola® Company sponsorships, and a marquee project in the works for SeaTac Airport. Jones was also planning to become involved with those planning a large scoring system for the Kingdome. All this plus just getting himself and his wife settled and getting the office up and running meant that Jones had his work cut out for him.

He remembers those early years as exciting times. He was learning his way around a metropolitan community of millions after having grown up in a small South Dakota river town of about 400. He found Feil's experience, helpful nature and knowledge invaluable as he ventured out into the community. Likewise, Jones' efforts at customer relations and his enthusiasm for what he did impressed Feil. "Marlo was very knowledgeable and really worked hard to continue the business I had started years before," Feil said. "I had a very good working relationship with him and I tried to help him when he had questions for me. I never heard any complaints from any of my previous customers." Jones' good work in his pioneering effort for the company did not surprise Kurtenbach, either.

An early baseball scoreboard in Seattle with the Scoreboard Sales and Service logo (under the word "inning").

"Having people who would be able to both service the product as well as sell it and install it properly was important to us," Kurtenbach said. "Marlo fit that requirement very well. He had a positive outlook on life, had self confidence and was not afraid to walk into a room." In assigning that important and pioneering first sales and service office to Jones, Kurtenbach said he felt very confident Jones could carry out the marketing needed in that new area, "which was primarily attending trade shows at the state level where coaches, athletic directors, school administrators, school board members, and others associated with school management might be involved in purchase decisions." Looking back, Kurtenbach believes Jones did a terrific job "and was just a great individual to take on that challenge."

Jones also remembers vividly his first year in the northwest and the challenges of doing business in a community far from home. He describes some of his experiences in his own words:

"Starting out and for many years thereafter we serviced all brands of scoreboards, both electro-mechanical and solid state, usually at the customer site or at our office. At times, a module would need to be sent to the

Seattle Mariners scoreboard at the Kingdome.

manufacturer, but most of the time the parts were available and could be procured locally.

"Chet (Feil) was a great help to me the first year, offering good advice and assistance for installations and service. He had also worked with many of the customers over the years and understood how to sell scoreboards, helping me in that process, too. Many of the scoreboards we serviced were the older electro-mechanical Nissen-Scoremaster, Fair-Play and Naden, to name a few. When it became cost prohibitive to continue to repair them, we could offer a new scoreboard proposal."

Jones said Feil's good reputation as a competent and fair scoreboard technician was a big help. "His endorsement of me to his customers helped me earn their trust. All I had to do was continue to provide good service, follow-up and help the customer with their needs."

"Early on I remember Al telling me that 'selling is just helping people' and this is still true today. However, it is ever more important today to understand or discover what help the person wants before determining which solution is best. The vast product selection we have now with a one hundred page catalog is significantly larger than our eight-page catalog we had in place at that time," Jones said.

Marlo at a trade show.

His first year impressed upon him that he would be a jack of all trades, especially in the scoreboard repair area. He not only kept Daktronics equipment in tip-top shape and running but was also called upon to repair and maintain Fairplay, Nevco, Scoretyme, Nissen Scoremaster, All American, American Sign and Indicator, and Daric, among others. Feil was of great help in working on these displays, and some of his testing equipment was used, as well. Jones at times also had to devise his own test equipment and also sometimes used a scoreboard trade-in that would work as a test circuit. Jones learned that some of the parts were interchangeable between the different manufacturers with minor treatment. "When a test fixture was unavailable," he recalled, "the only choice was to repair the problem on-site."

Jones still remembers his first service call repair order. It was for Seattle Pacific University's Daktronics Model BB-1718 scoreboards. He must have done a good job then and on subsequent calls. The university remains a loyal Daktronics customer today, Jones says proudly. His records from that first year show that in the first month of operation, he had three service calls to four-year universities in the area, one call to a community college, six to high schools, two to junior high schools, and one to a boys and girls club. Six of the locations had Daktronics scoreboards, three had Naden, one a Nissen Scoremaster, two had Fair-Play equipment, and one had a Scoretyme. "Gradually over the years," he said, "we were able to replace most of these with Daktronics scoreboards which became the standard for most facilities in the area."

The proliferation of Daktronics scoreboards was partly because when Jones arrived in Seattle, there were three active Daktronics resellers/dealers. In addition to Chet Feil's business, the other resellers were NorPac, which sold primarily through the new construction market, and Learning World, a seller of school supply products throughout the Northwest and Alaska. "Because of this activity," Jones said, "we had always enjoyed a good reputation for service on our products."

Test scoreboards in the Seattle office (1988).

Seattle test bench.

Daktronics scoreboards became the standard for most facilities in the area. This is the Bellevue High School football scoreboard in Seattle.

Jones did have some hurdles to clear during the early years. In 1988, Daktronics scoreboards were ETL listed, and the State of Washington recognized this as an equal for UL. "But some of the local jurisdictions didn't accept it," Jones said, noting that Nevco was a UL listed product "and this made it difficult sometimes to sell to a customer that needed an easy reason to continue to buy Nevco."

Jones credits the work of Al VanBemmel, a Daktronics scoreboard design supervisor, for assisting him with documentation that helped explain to prospective customers the ETL versus UL (Underwriters Laboratory) rationale. He recalled that often, even that was not sufficient for the potential buyer. "But at least we did make Nevco work for the business they secured."

With the help of Feil, and with Jones' persistence and hard work, business improved. By fiscal year 1990, sales of new product was $10,000 a month, nearly double Jones' projected figure. Service revenue that year was about $2,500 a month, but "we became better," Jones said. Within a few years, service revenue in Jones' shop reached $200,000 a year. Speaking of the revenue from service, Jones remembers that a sign company owner he met during his installation years told him it was important to get enough service agreements in place to cover payroll. "It was good advice and for many years we did that," he said.

Things were falling in place for Jones and for Daktronics as the Seattle office matured, continued the good reputation instilled by Feil, and grew in customer contacts. The Daktronics "service first" premise and motto upon which the first scoreboard sales and service offices were based was proving to be very effective. Jones was tireless in his pursuit of making new friends and informing them of his willingness to help service their wares, working his way through boxes of business cards and encouraging everyone to call him at any hour of any

Daktronics scoreboard design supervisor, Al VanBemmel.

day if help was needed. He was getting his adventurous foot in the doors throughout the Seattle area.

Both Jones and Daktronics were learning how to make a remote sales and service business function, and the unique to the industry "service first" approach was effective in opening those doors. Meanwhile, back in the home office in Brookings, others were taking note, too, and augmenting what was being learned in real-world situations in Seattle with other lessons that would be invaluable as the company continued to work toward its goal of opening four new scoreboard sales and service offices a year.

Among the many lessons learned was that Seattle was a good classroom setting for future Daktronics scoreboard sales and service office managers, and that Jones was a good teacher.

3 Getting Started

The experiences and on-the-job lessons learned by Marlo Jones in establishing Daktronics' first satellite Scoreboard Sales and Service office in Seattle in 1988 were invaluable in setting the course for future offices being planned. More "how to" examples were acquired during the experiences Ed Weninger and his compatriots in the earlier attempt with sales and service offices in Illinois and other Midwestern states.

Carla Gatzke took all of this and some practical solutions as well to author an operational manual as a guide for those launching new sales and service outposts. With the company poised to launch this national network, her document, completed in April of 1990, was the blueprint for those who would follow in Jones' footsteps, establishing scoreboard sales and service offices in other markets across the country.

Part of Gatzke's work included the office start-up lessons learned by Weninger in DeKalb, Illinois, and the others sent out in the firm's first foray into the world of satellite offices. Co-founder Kurtenbach, who is Carla Gatzke's father, considers Weninger's five years in a somewhat makeshift office operating on a knotted and frayed shoestring in Illinois in the 1970s as a company bellwether. It was breaking new ground and leading the way. "The experience and background Ed Weninger brought back to us from his pioneering effort in Illinois was essential in helping first identify and then train technicians who would have a high probability of success in our future scoreboard sales and service endeavors," Kurtenbach said.

Gatzke's meticulously written document was also of value for those young employees who would be identified for satellite office leadership.

Nothing in the way of opening or operating a satellite scoreboard sales and service office was overlooked. It even included the number of file folders that would be needed to help get a new office up and running (one pack-

---

2.7 Van

Additional test equipment, supplies and inventory is to be located in the company van. Note that the office should have a garage so that equipment can always be stored safely in the van. If a garage is not available, an alarm on the van is necessary. However, damage can still be incurred by the van and equipment stolen even if an alarm is installed, so a garage is preferable.

The van should have a ball on the hitch so that rented trailers can be attached for hauling scoreboards and other equipment. The following general equipment should be included in the van:

1. A complete map - like the Thomas Guide
2. Tire chains
3. Yellow pages
4. Flashlight
5. Boots
6. Fire Extinguisher
7. Consider Cellular phone

The following spare parts should be included:

1. A competitor control console like the Nevco MPC3.
2. Daktronics All-Sport 2000
3. Daktronics All-Sport 2200
4. Fairplay 60F driver
5. Daktronics blue control console
6. Daktronics empty blue control console carrier
7. Daktronics ST213 Shot Clock with test display

An installation box in the van should include the following equipment that is valuable during installation:

1. Hammer drill
2. Socket set
3. Drill bits
4. Hammer
5. Assorted nuts and bolts
6. Lag bolts and shields
7. Cable ties
8. Dry wall screws
9. One Daktronics indoor driver
10. One Daktronics outdoor driver

7

Carla Gatzke's Scoreboard Sales and Service startup manual. This section shows the service van inventory (continued on next two pages).

The following general equipment should be included in the van:

1. Ladder - preferably a wing 22 foot folding ladder.
2. Wire and cable, 20 conductor shot clock cable and Daktronics W1077 cable.
3. Lamps
4. Block and tackle for lifting scoreboards.
5. Guide rope for lifting scoreboards.
6. Hacksaw
7. Extension cords
8. Set of knockouts (preferred)
9. Y-adapter cable (two each)
10. Coaxial cable (two each)

The van should also include a parts kit and the general tool box.

Two parts caddies are to be included in the van. The parts caddies are fishing tackle boxes, with spare parts to be used during service. Parts caddy one includes:

1. Daktronics connectors
2. Competitor connectors
3. Daktronics switches
4. Competitor switches
5. All-Sport 2000 jacks
6. Small sockets
7. Daktronics and competitor pins and pin popper

Parts caddy two includes:

1. Daktronics and competitor fuses
2. Integrated circuits
3. Triax
4. Competitor lamp/lenses
5. Daktronics and competitor resistors
6. Miscellaneous wire, cable ties, and wire nuts
7. Daktronics and competitor lamp sockets

The tool box includes tools used both in the field and at the repair bench. The tool box includes the following:

1. Battery operated drill; one spare battery
2. Drill index
3. Vice grips
4. 8" Crescent wrench
5. Small channel lock
6. Amp 22-0 1-9 W Crimper
7. Amp VI 90277
8. Two soldering irons
9. Solder sucker
10. 6" Slip joint pliers
11. Nut driver set

Service van inventory from Carla Gatzke.

```
12.  Large side cutters
13.  Small side cutters
14.  Small needle nose pliers
15.  Wire strippers
16.  Two pin poppers
17.  Five screwdrivers
18.  Digital meter
19.  Hand metal nibbler
20.  Draw knife
21.  3/8 Socket set
22.  Miscellaneous components (crimps)
23.  Tool box
24.  Flashlight
25.  Pop rivet tool
26.  Hammer
27.  12" Aligning punch
28.  25' Tape measure
29.  Combination wrench
```

Service van inventory from Carla Gatzke manual.

age). The booklet suggested, for example, work bench Ps and Qs (or rather its Ws and Ds) recommending that a can of WD-40 be among the items for the office shop. And new managers were reminded not to forget to have a fire extinguisher within easy reach.

Gatzke, who is now Daktronics' Vice President for Human Resources, also listed the various report and sales order forms, repair orders, and other paperwork a new office manager would need. She outlined the step-by-step details for mass mailings and her document suggested how to keep records on lead tracking. It was all there, everything one would ever want to know about managing a Scoreboard Sales and Service office but was afraid to ask. There were even lists of what basic equipment should be carried in the office service truck or van. For example, there should be "two parts caddies" and a tool box, containing—among other things—two soldering irons and one small channel lock rattling around in the service truck.

In all, Gatzke wrote an impressive 32-page manual remarkable for its detail and inclusion of all possible contingencies. Nothing was left to chance. There were even tips for customer contacts and how to deal with people. ("It is much easier for a prospect to give an order to a person that they know, than to give an order to Scoreboard Sales and Service.") Al-

Seattle inventory (1988).

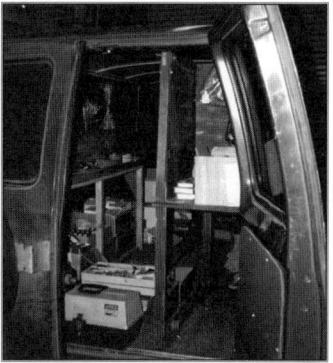
The back of a Seattle service van.

though the document has been edited and changed and has grown in size over the years as new products came on line making some work procedures obsolete, the basic outline is intact and is always within reach of new Daktronics managers just starting up.

Jones, out in the first Scoreboard Sales and Service office in Seattle, was impressed with the manual as he worked out the kinks and figured out the secrets of organizing his new office in Seattle. "It was a great tool for opening a new office and also provided some standards by which to operate much like a franchise," he said. "Many of the procedures and processes remain the same today, although many have been converted to an electronic format."

As Jones gained confidence, knowledge, and experience in Seattle, it became his office that was charged with providing on-the-job training for others heading out to establish new regional offices elsewhere. Backing up these introductory field orientation efforts in Seattle was a seasoned team at corporate headquarters in Brookings tasked with organizing and conducting manager training programs and assisting the satellite office in every way. The team leadership included Frank Kurtenbach, Gary Gramm, and Lynette Smith, and they often consulted Ed Weninger. During the early years, the satellite offices initially consisted of either one or two persons depending on the training schedule of the office manager selected for the program. The major administrative activities of the scoreboard sales and service efforts came under Gramm's responsibility, and he continues in that role today, with overall responsibility for the High School Parks and Recreation market.

309 S. Cloverdale St., #C-2   Seattle, WA 98108
Phone (206) 763-6434

February 20, 1991

Al Kurtenbach, Pres.
Daktronics Inc.
PO Box 128
Brookings, SD 57006

Dear Al:

    Al, I'm sorry this is a little late. I have been very busy trying to keep the parenthesis from appearing on the bottom line of my statement.

    As you will note from my plan, things look very good for finishing the year with a profit. January & February were both good months, so if I can generate something for March & April, we will be in good shape.

    If you have any questions regarding this information, please call me. Also, if you have anything to add, I welcome your comments. Thanks for your help this past year.

Scoreboard Sales & Service

Marlo J Jones, Mgr

Marlo Jones's letter sent to Al Kurtenbach.

SCOREBOARD SALES & SERVICE™

309 S. Cloverdale St., #C-2　　Seattle, WA 98108
Phone (206) 763-6434

PLAN FOR
SCOREBOARD SALES & SERVICE/SEATTLE
JAN. 91' THRU APR. 91'

SUMMARY OF May thru Dec. 1990

1. SALES ---$ 60,363
2. SERVICE -$ 35,414
3. GROSS PROFIT -------------32.3%   OR $ 30,902
   NET PROFIT   -------------12.0%   OR $ 11,523

GOALS FOR JAN thru APR. 1991

1. SALES ---$ 8,000/MO. OR $ 32,000
2. SERVICE -$ 3,000/MO. OR $ 12,000
3. GROSS PROFIT ---------------------25% OR $ 11,000
   NET PROFIT   --------   ----------10% OR $ 4,400

　　January and February have both been good months so I expect to exceed the monthly goal in each them.

MARKETING PLAN FOR JAN. thru APR. 1991

　　1. Attend Conferences:
　　　　March 21 & 22
　　　　Washington Activity Coord. Conference   Pasco, WA

　　　　March 30th
　　　　Target for Summer service and Pre-AD conference
　　　　Mailing. (try to tie WSU into it)

　　　　April 24, 25, 26
　　　　Secondary School Athletic Admin. Conf.   Moscow, ID
　　　　(Tie WSU, Ed K., & state of Idaho into conference)

　　2. Continue to call on schools and people while on service calls.

　　3. Start Data Base of Electrical Contractors in the area and mail to them regarding baseball scoreboards. Also update architect's files with new FB specifications. This gives me a reason to call on them.

Signed: _[signature]_ Date _Feb 20, 1991_

Marlo Jones's January 1991 through April 1991 Business Plan sent to Al Kurtenbach.

In 2006, Daktronics realigned to a business unit structure, and the small scoreboard venues came into that scoreboard business unit which is managed now by Dan Bierschbach, Vice President for Schools and Theaters, and after sales service with Gramm and Weninger reporting to him.

One of the early methods of maintaining regular contact with Jones and the small cadre of sales and services office managers was a regular monthly telephone conference. During these sessions, corporate headquarters briefed office managers around the nation on what was new or changing in the industry and at the Brookings campus. Managers could likewise commiserate and make suggestions for changes that would be beneficial from their in-the-trenches point of view. The conferences were valuable because of the sharing of experiences that often uncovered challenges that many of the managers occasionally faced. The first official conference call of this type was made on August 17, 1992. These regularly scheduled calls continue today, although with the increased number of offices and personnel in satellite offices, the procedure has been structured for small groups with similar responsibilities.

These conference calls, the annual seminars at the Brookings campus, and the general assistance provided by staff in Brookings has proven very helpful for all concerned, with office managers learning from corporate and corporate learning from the experiences of office managers. The give-and-take runs the gamut, from periodic visits by Gramm or other company officials to help in special situations or in making new product presentations to such things as coming up with the first Scoreboard Sales and Service logo that was developed by Mark Steinkamp, Daktronics marketing and sales support manager.

Jones, who was by then well established in growing his Seattle operation, remembered that he felt somewhat alone and adrift during the first several months of solo work until the conference calls were initiated. He tried to make the best of it by seeking out his peers in the Seattle area. To help him become better at what he was doing, he also enrolled in night school and took weekend classes and eventually earned an MBA degree from City University in Bellevue, Washington.

When Jones arrived in Seattle, competition in the area by other companies was significant. One of Jones' promotion ideas, which he felt Daktronics needed in order to compete with the Fair-Play Company's five-year lamp warranty, was very successful. "I wanted to start a free lamps program to all customers who had our basketball scoreboards that used the #656 wedge base lamp," Jones said. He was aware that the #656 was a good lamp that seldom burned out so costs would be minimal. He presented his idea to Al Kurtenbach, who asked for supporting documentation on the plan before giving the go-ahead.

By fiscal year 1992 Kurtenbach had the information Jones had researched and written. He accepted the free lamp proposal provided the customer understood that the free lamps did not include labor as part of the offer and that the plan included only the more bulb-friendly indoor scoreboards. "The program was very effective in getting our customers to contact us for small lamp quantities needed to maintain the excellent appearance of their dot digit scoreboards," Jones said. Daktronics mailed customers the lamps in quantities of ten to a box that was clearly labeled "Free Replacement Lamps." Jones said that the program was very successful and helpful as a public relations goodwill gesture that was appreciated by customers for many years until the LED scoreboard rendered incandescent lamps more or less obsolete.

Seattle test scoreboards (1988).

Introducing Daktronics in the Seattle area was a challenge for Jones, as it would be for other mangers who would open other scoreboard sales and service offices throughout the United States. All of those first office managers say that

```
                    SCOREBOARD SALES & SERVICE
        POLICY:    FREE REPLACEMENT LAMPS TO USERS OF INDOOR BOARDS
                   USING THE #656 LAMP. i.e. Current standard BB & ST.
        GOAL:      TO HAVE PEOPLE REMEMBER THAT WHEN THEY NEED SCOREBOARDS,
                   OR SERVICE, BE SURE TO CALL US.

   Benefits for the customer:
        1.  You receive something extra when buying a board from me.
            You don't have to worry about finding a place to buy lamps
            or whether you can get them 5 cents cheaper someplace else.
        2.  You get a feeling of "this company wants to take care of
            me". Even more, they have confidence in their product.
        3.  You are more likely to keep your scoreboard looking good.
            Other teams will see it and comment on it. You may tell
            them why it looks so good.
        4.  You realize the company will take care of you after the
            sale, and not forget about you. This makes you feel
            comfortable with who you are dealing with.

   BENEFITS FOR SCOREBOARD SALES & SERVICE:
        1.  We get referrals from our existing customer base due to
            the fact that our boards look good, and we take care of the
            customer.
        2.  Every call for lamps is a sales opportunity. They may need
            a protective screen if they are using a lot of lamps, a
            second scoreboard for the other end of the gym, an outdoor
            board, a portable, service, or a spare control console.
        3.  We increase service orders since we are in contact with the
            customer. He may have a problem that he doesn't know how to
            fix and has just lived with it. Or, his BA or FB board may
            have a problem.
        4.  We increase sales orders for small items such as remote SS
            switches, carry cases, and shot clocks & other options.
        5.  We get name recognition every time the client needs lamps.
            Especially when the board is getting up in age.
        6.  We use it as a selling point. We have an excellent product
            and are willing to stand behind it. We will do our part if
            the customer does his part "TO REMEMBER OUR NAME".
```

Proposal for "Free Replacement Lamps," submitted to Al Kurtenbach by Marlo Jones in 1992.

they found that Daktronics was not exactly a household word hundreds of miles from Brookings.

Jones remembered his experiences in working with Seattle area architects in the late 1980s. He and others would learn of new schools going up and the need for scoreboards from athletic directors at trade shows,

```
                        Operations Strategy

            Do this as a policy of ours. Withhold the right to cancel
            the program at any time. And sell it as that.
              'Our policy is to give out replacement lamps for your new
            scoreboard'.

            Require the bad lamps returned if something doesn't seem
            right.

            Lamp technology has only gotten better so life expectancy
            should increase as time passes.

            I haven't sold 150 lamps in the two years I've been here.

            Think of it as a cost of marketing. Similar to a conference
            or mailing. We get more out of a dollar spent, since each
            5 or 10 lamps is a quality contact for the year. At a
            conference, I spend $300 to talk to 100 people and get
            20-25 leads.

 CALCULATIONS: (Average Scoreboard)

     2 Hrs/Game X 10 Games/wk = 20 Hrs/wk.

     20 Hrs X 28 Wks/Yr = 560 Hrs/Yr.

     (Lamp Life) 10,000 Hrs/560 Hrs = 17.86 Yrs.

     17 digits X 20 Lamps = 340 Lamps/scoreboard.

     340 lamps / 17.86 Yrs = 19.04 Lamps/Yr.

     19 lamps X $.28 = $5.32

     Total cost per year, per scoreboard is $5.32 + postage.
```

Proposal for "Free Replacement Lamps" continued.

conventions, and during other contacts. "My task then was to contact the architect and try to persuade the design firm to at least allow Daktronics to bid the project by approving us an equal product," he said. At the time, "equal" usually meant "as compared to Nevco because it was the standard that architects used," Jones said. This was because Nevco was easy and simple to specify for these installations because it did not regularly change the major items related to a model, such as overall size, digit size, digit style or other physical features, "and the product seemed to work," Jones said.

At that time, another perceived Nevco advantage was that the control console was advertised as interchangeable between different models of scoreboards. "In some cases that was true," Jones said, "but not universally so." This all helped make it rather simple for the architects to specify Nevco year after year, and also for them to justify the value in having basketball and football scoreboards that could use the same control consoles interchangeably by just changing the overlay and inserting a new code at start-up.

Jones considered this affinity by architects to one product a disadvantage for the user because the scoreboards tended to look the same for generations, and replacing a scoreboard rarely improved the image of the gymnasium or field. This reluctance to change, Jones said, and the fact that the scoreboards used basically a variation of painted 120 volt "Christmas tree light bulbs" were major reasons we were able to successfully get products specified. "The architects would listen to the owner's requests when it came to what scoreboards they wanted for their facilities," Jones said, "so with consultative selling we were able to greatly reduce Nevco's market share relating to new construction."

Another innovative effort Jones helped bring about was to increase Daktronics' distribution for the new construction market. Construction projects are designed using a very systematic process, and each trade or type of work is grouped in a division. "Prior to 1988, NorPac had the majority of our new construction business in Washington as well as in Alaska," Jones said.

NorPac bid division eleven products (a construction project's various stages are assigned a division number) to the general contractors because their specialty was to provide products that might fall into that group, such as baskets, scoreboards and wall padding. "The challenge with this was that many times the scoreboards, especially outdoor scoreboards, were specified by the electrical engineering firms working with architectural firms, and were in division sixteen rather than eleven. Later on, when electronic readerboards became popular, division ten (signs) became a factor as well. This meant we had to have a system in place to bid projects directly, and I found when Nevco or another manufacturer bid directly, we had to have a way to compete with that. After getting the products approved or specified, we found the product consolidation companies would rather buy from us because we had the approved product, support and after the sale service."

From Jones' Seattle experiences, he also became a company expert of sorts in dealing with scoreboard sponsorship by various companies, and he was often called upon at Daktronics seminars to talk about competing in this marketplace. "One thing that would come up periodically would be the subject of the end user, Coca Cola® and Pepsi Cola®, and how to sell without upsetting one or the other party." He said that when he started in Seattle, "I guess I didn't know any better and I always assumed I was in the scoreboard business and they were in the soda business." Jones said he believed that "if we could help the bottlers with their customer requirements without restricting our ability to market our products directly, we would be better able to sell our product."

Jones stressed that he was always careful to treat all of the parties fairly and with the same level of high quality service. By doing this he was able to work with all the bottlers simultaneously. "I think the bottlers appreciated the fact that the schools preferred the same scoreboard brand because then they knew what the other was paying for the scoreboard and it removed a previously unknown variable for them," Jones said. "I can believe it was a challenge when one of the bottlers was willing to put their advertisement panel on a cheaper, low quality product," he said. "This would require the others to either reduce their equipment standards or dip into the margins on their products to make up the difference."

He said that with his success with Coca Cola® and Pepsi® locally, he was convinced he could offer that same full service solution to all of the contractors and school supply resellers in the area, especially when the end user preferred Daktronics products and service offerings. "When working with just one reseller, it was always a gamble when partnering with one because of their other product offerings and the risk was too great," he said. "Market share using an exclusive method at least in this area would have been twenty-five percent or less using this distribution method. Therefore, we have always maintained the right to bid or quote to anyone, including the end user," Jones said.

Tacoma Raniers scoreboard (1994).

He noted that he learned that the architects and schools liked this approach because they could specify and hold to the specifications when there are multiple providers of the product. The resellers' package products together and by having a source for the products specified in a bid specification gives them more opportunities to win jobs, so they also like the ability to work with us on more projects. "The goal, "Jones said, "is to have people want to work with us because of our product quality and its appearance, as well as our service, rather than to have to work with us because we have some sort of exclusive arrangement." Jones noted that because of the fairly recent tendency for schools to move away from in-school vending of some products such as soda, he is finding that the dairy industry and other sponsors are filling the void.

With all of the challenges that faced Jones and the other office managers as more joined the growing list of company sales and service offices, there was often little time for working out the details of starting a new office, such as what equipment and supplies were needed, how to find a good office location, dealing with the landlord and doing just the normal, day-to-day things that would free up the manager and others for the business of growing the office.

The young employees leaving for office managers' jobs in larger population areas were barely out of college. While willing to work and do what it took to succeed, they might be naïve in some ways, and in need of central office backup to help them get started. It was soon determined by the team of Frank Kurtenbach, Gary Gramm and Lynette Smith that the goal of setting up four new offices a year would be less arduous for each new office manager if someone other than the manager was there to worry about such details. "Just going out into a new territory to open a new office and then pioneer the development of the territory could be a very lonely and sometimes discouraging activity," Al Kurtenbach said. Lynette Smith was tasked to be the one to assist in on-site office establishment and to handle the details and logistics of getting the office functional. "She turned out to be a wonderful person for doing that," Kurtenbach recalled. Gramm, for whom she worked, agreed. "She did it all, and she did it well," he said.

Smith not only coordinated the details for establishing the new office, but she was almost a sort of mother figure for those mostly young men who would go out and open new offices. "After we started to appreciate more the importance of the need for a liaison person, Lynette traveled to the site and really set up the internal workings of the new office," Kurtenbach said. She would coordinate furniture purchases and insure the telephone functioned properly. She set up the filing system so that these tasks were not taking up the time of the office manager, who would then be free to concentrate on the initial efforts of getting out into the community, knocking on doors, and talking to schools about their scoreboards and inventory that was currently in use in the community.

The selection of satellite scoreboard sales and service office locations didn't depend on darts thrown at a map of the United States. They

Lynette Smith at a Daktronics picnic at the Kurtenbach farm.

were picked only after thorough study. Often, the location would match up with a large venue customer, such as a professional sports team. Other times, a site was selected to be better able to compete with others in the same business.

At the time the new offices were being opened, Daktronics still had some independent dealers so Daktronics would scout out areas that were not in too close a proximity to successful independent dealers, usually in a state or area in which Daktronics was not generating a good deal of business.

Once an area was targeted, a central location within it was selected. This was usually a larger community centrally situated. If possible, a sales and service site would be within easy access to a major highway system or an Interstate highway, and it would be reasonably close to an airport. Having a technical college close by was also helpful, and an important factor in attracting technical students as interns to assist with service and to develop into potential full-time technical employees. From that central sales and service office, routes to outlying areas were determined and mapped out for maximum travel efficiency as the technical and sales staffs at the office were increased year to year.

Early LED basketball and shot clock in Seattle.

Based in part on the lessons learned at the Seattle office, a mission statement for the scoreboard sales and service locations was outlined and written by Kurtenbach and Gramm. The scoreboard sales and service locations would have an "effective area" that included a two-hour drive radius. Within that so-called "effective area," the office would service all makes of scoreboards in kindergarten through twelfth grade schools, boys and girls clubs, parks and recreation departments, and junior colleges. The office would also sell Daktronics product to those entities and service all Daktronics products in that territory. The overall mission of the office was to support Daktronics installations and training, working with project managers and markets. And finally, the mission statement said that sales and service offices would "seek other opportunities to sell and service."

While the remote offices usually were launched with only an office manager-technician on staff, a second person might be hired as the workload increased. That person would typically be a part-time office coordinator who would staff the office while the technician was out servicing, selling, and making contacts. Over time, the office coordinator came to be considered an integral member of the team. He or she would become knowledgeable about various Daktronics products and procedures. The office coordinator would be trained on following up on quotes and gathering pertinent data on service problems that might easily be solved by contacting the help line at the Brookings office.

A tried and true model for sales and service offices has slowly evolved. Its core is a three-part plan that assumes the market share in both sports and commercial markets at the time of office launching would be somewhere from zero to ten percent. As the office added staff and experience over time, the eventual market share in this model was estimated to ultimately grow to seventy-plus percent of the market.

During stage one of scoreboard sales and service office development, the model assumed the staff would consist of one salesperson/technician with about an equal balance of the technician's time spent in providing services and twenty percent of time invested in sales. The service area in this initial start-up phase was established to be within two hours driving time of the office.

Progression to phase two included the addition of staff, and an expanded service and sales area beyond the approximately two hours from office limit. Longer routes branching out beyond the two hours radius are mapped out, with possibly four planned. Because of the added distance and the increased number of destinations in this new configuration, two days of staff time might be required to cover each route.

The third and final phase of the Daktronics sales and service model became operational when about eighty percent of staff time was devoted to sales, and service response required about twenty percent of the total staff time. Of course, in the third stage, the sales and service staff was assumed to have grown to exceed five salespeople/technicians. With that added staff, the office was better able to guarantee same-day service calls within two hours of the office. And added to the four routes radiating out beyond the two hour office radius in stage two would be a number of longer routes that would require up to a week to cover.

Of course, there would always be variations of this model because of local conditions and characteristics. But the model has been and remains a starting point that serves the company well. Today, Daktronics has forty-nine offices strategically placed around the nation. The office functions have been expanded beyond scoreboard sales and service endeavors. They now support the litany of Daktronics products and include full-time sales and service people, as well as staff serving the large sport and live event venues, the commercial dealers, and the transportation customers. Many current sales and service offices have a variety of talents to meet every contingency or challenge. "The general idea now," said Kurtenbach, "is that whenever possible we would have these people work out of the same office so there is more opportunity for mutual support as well as backup where one technician in a given market might help a technician in another market because of the urgency that exists at that particular time." Of course, they interact and work as a Daktronics team to serve the assigned geographic area.

Today, Jones' responsibilities in Seattle have grown beyond what he ever envisioned as a young college graduate starting out. Kyle Williams now manages the Seattle office and Jones is High School, Parks and Recreation Region Manager for California and the Rocky Mountain regions. States in the Rocky Mountain Region are Montana, Wyoming, Colorado, Arizona, New Mexico, Idaho and Utah. A licensed Journeyman Maintenance Electrician and holding an Electrical Administrators license, Jones serves on the Electrical Engineering Technology Advisory Board of Western Washington University and is also a member of the American Marketing Association. Says co-founder Aelred Kurtenbach, who reached out to him in 1988: "He's doing a tremendous job out there in the field for us both in managing his regions and serving as a leader as we gravitated toward a more regionally-oriented sales force."

So with Jones growing the Daktronics presence in Seattle, notching the trail and helping introduce future regional office managers and regional leaders, the Daktronics' burgeoning sales and service network was on its way. Following in Jones' footsteps in the formative years when everyone was learning about how best to meet the scoreboard sales and service challenges, were others like Darrell Thiner in Indianapolis, Dave Marsh, who launched the Billings, Missouri, office in 1990, Bryan Nagel who started up the Baltimore, Maryland, office in 1991, Perry Grave, who journeyed to Oklahoma City, Paul Wildeman in San Antonio, and Tim Branton, who pioneered the Lexington, South Carolina, office in 1993.

Meanwhile, back in Seattle, Chet Feil, whose 1988 decision to sell his scoreboard parts and service business to Daktronics, has quietly retired, but continues to be active in his church and is presently also the designated, official digital photographer at the senior apartment complex in which he and his wife Ruth now reside. Jones is forever grateful for the assistance Feil provided in the early years. "I'm sure if I ask him to even today he'd drop everything and give me a hand," Jones said.

4
HALL PASSES AND THE SECRET SERVICE

Darrell Thiner, who was selected to manage a Daktronics' Scoreboard Sales and Service office, wasn't yet twenty-five-years old when he started making introductory visits to high schools in the Indianapolis area. His goal was five or six schools a day.

The 1990 graduate of South Dakota State University with a degree in Electronics Engineering Technology was somewhat taken aback when, while strolling down one school's hallway, he was stopped by a teacher-hall monitor who asked to see his student hall pass. He apparently looked young enough to be a high school student. That didn't deter him from performing like a veteran in establishing from scratch the first Daktronics Scoreboard Sales and Service office in the nation, after Jones' experiences in building upon the business in Seattle started by Chet Feil.

"We put a lot of faith then in young men like Darrell and others," said High School/Parks and Recreation (HSPR) market manager Gary Gramm. "We had faith that they had the common sense and training and desire to do the job and that, on their own, they would do what is necessary."

Darrell Thiner in the Indianapolis, Indiana, Daktronics Sales and Service office.

Thiner is a native of Windom, Minnesota, where he played on his high school's two-time state championship basketball team. While in college, he took a fall semester off from studies to work as a commercial intern at Daktronics. Among his assignments was traveling the country demonstrating the "Glow Cube" display until December of 1988. He said that his experience "towing around the trailer and the 40 x 128 Glow Cube demonstration to potential customers" was exciting and gave him experience in selling and in customer relations.

He then worked in Customer Service repairing addressable drivers, the Olympian 80 message center controller, and the "SuperBrain" message center controller. Thiner also manned the Help Desk for a time. He remembers that it then consisted of two telephones, a pencil, a notepad, and a platoon of filing cabinets. Before his graduation in December, Gary Gramm asked him if he'd be interested in opening a second Scoreboard Sales and Service office. Gramm explained that it would be fashioned after the Seattle office then being developed by Jones, "but this one would be started from scratch." Gramm and Kurtenbach had their sights set on an office in either Columbus, Ohio, or Indianapolis, Indiana.

Gramm and Thiner flew east to check out each potential city. Afterwards, Thiner voted for Indianapolis and Gramm agreed. Back in Brookings, Thiner started working full time for Gramm in the HSPR market. He was learning the ropes, and in March he was assigned to man a Daktronics booth at the Indiana Athletic Directors convention in Indianapolis, his future home base. "I'd attended a couple of national HSPR-staffed conventions before with Gary Gramm, but this was my first convention alone," Thiner remembered. He confessed that it wasn't exactly a busy time for him. "No one knew me or had ever heard of Daktronics," he said. Meanwhile, his two competitors at the convention had "plenty of traffic," he recalled.

After that initial and somewhat disappointing experience in Indianapolis, Thiner flew to Seattle to spend time with Marlo Jones. The two went out on service calls, visited a new construction company office, and installed a new Glow Cube scoreboard at the Kingdome that displayed out-of-town scores for the Seattle Mariners. "That week with Marlo showed me I would be wearing many hats and be the face of Daktronics to people who had never heard of our company," Thiner said.

By May, Thiner and his wife, Paige, who had graduated from Dakota State College the week before, flew to Indianapolis to find an apartment and, with Gramm's help, to locate a suitable office for the new business in town. As luck would have it, they picked Memorial Day to move into their new apartment. It was just seven miles from the Indianapolis 500 racetrack and the Thiners found themselves battling 200,000 people and unbelievable traffic in town for the big race.

Thiner wasn't deterred. He immediately went to work getting to know the territory and introducing himself to potential customers. He said opening that new Indiana office gave him some idea of the difficulties and what it must have been like for Al Kurtenbach and Duane Sander when they started Daktronics in 1968. "At first the phone didn't ring and I had trouble finding anyone who had ever heard of Daktronics," he said. At the time, Daktronics had only a few dozen high school installations in Indiana. The only large installation was in the Hoosier Dome. "When I introduced myself to athletic directors, they thought Daktronics was the DAK electronics catalog they received in the mail," he said.

Darrell Thiner at the work bench.

During his rounds, visiting schools and other businesses, many potential clients would ask if he could fix their scoreboards. "I'd say yes, whether I had ever seen the brand before or not." His approach was to "just jump in and try to figure it out based on my schooling and knowledge of Daktronics equipment." If he had trouble solving the problem or if he needed parts, he'd give Marlo Jones in Seattle a call, or he'd call one of the Daktronics dealers who repaired scoreboards, such as Chuck Hurley of Massachusetts.

Later in that summer of 1990, Thiner helped install one of Daktronics' first three-inch colored lens displays at Purdue University. "I welcomed that chance to work on Daktronics equipment since there was so little of it installed in Indiana," he said. Daktronics president Al Kurtenbach, who earned his doctorate degree in electrical engineering from Purdue, decided to take in the first Purdue football game of the season where the Daktronics display would be put through its paces. Kurtenbach stopped in Indianapolis to pick up Thiner's wife Paige, and they drove to West Lafayette where they met Thiner. The three of them enjoyed the football game, and especially the colorful Daktronics display that was part of the entertainment.

The next month, Daktronics national Sales Manager Frank Kurtenbach invited Thiner and his wife to attend a Monday night football game at the Hoosier Dome where the Colts would be playing. In the second quarter, to the horror of both Kurtenbach and Thiner, the Daktronics scoreboards in the dome malfunctioned. "The scoreboards started flashing and the horns would blast intermittently," Thiner remembered. The game was halted and the raucous Monday night crowd expressed its displeasure. Television cameras zoomed in on the board's Daktronics logo. "Frank looked at me and said, 'Thiner, let's go fix 'em'," Thiner said. Problem was, neither Thiner nor Kurtenbach knew where the scoreboard control room was located.

Thiner said he slowly stood up from his seat and nonchalantly removed his Daktronics jacket. "I suddenly felt very hot," he joked. The repair mission of Thiner and Kurtenbach was further complicated because Indiana native, Vice President Dan Quayle, was at the game. Security was especially suspicious of someone searching for the venue's control room. They tried to enter the press box to inquire where the control room was located but were turned back by the officious Secret Service crew that was talking quietly into the collars of their ubiquitous trench coats.

By this time, play had resumed and the referee had manually timed the remainder of the second quarter. Thankfully, it was halftime. "We were finally told that the scoreboards were controlled from the top row of the Dome, and we were allowed access to that room," Thiner said. But by the time they arrived at the control room, the operators had plugged in the backup controller and the third quarter was underway. "This was my introduction to the Hoosier Dome," Thiner said. Ironically, through all the succeeding years that he would be present at Hoosier Dome events, the Daktronics scoreboard has worked without any malfunction. It never failed again. To which Thiner remarked: "Thank goodness!"

Scott Dieck and Darrell Thiner

After a few years, Thiner got help in his growing Indianapolis office from Scott Dieck, who in February 1998, became office manager of the Scoreboard Sales and Service office in Lexington, South Carolina, and from there was transferred to the position of office manager in Denver, Colorado. Thiner left to establish an office in Ankeny, Iowa. Succeeding Thiner in Indianapolis was John Paloma, who is now the manager of the Daktronics Sales and Service office there.

Scott Dieck testing a scoreboard in Indianapolis.

Because in the 1990s there were few Daktronics employees in the eastern United States, Thiner was often called upon to help with large installations in that area, working on projects at Wright State in Dayton, the University of Kentucky, Ball State, University of Louisville and a host of other universities, plus work for the Indianapolis Indians and the Indianapolis Motor Speedway.

One of his most exciting moments while in Indianapolis was in February of 1993. A phone call from Jim Morgan in Brookings asked his help in an emergency situation brought on by a blizzard. An intern was driving a van with a large trailer to Connecticut for a demonstration to that state's Department of Transportation. During the storm in northern Indiana, a spot of ice on I-80 launched the van and trailer in a 180-degree spin. No damage or injury resulted, but the intern had wisely retreated to a nearby hotel to wait out the storm. Meeting the Connecticut demonstration time and date was critical, and Morgan asked if Thiner could drive to the hotel and help the intern in the long drive on to the demonstration date for Connecticut transportation officials. "We succeeded in driving out of the storm and arriving in time," Thiner said.

When Thiner arrived in Indiana to establish the new office, scoreboard sales in the area were near zero. When he left in 1999 to establish an office in Ankeny, Iowa, annual sales approached $600,000. He said the most important requirement needed in growing the Indianapolis office was "determination." He said he never gave up. "I was always there for my customers at all times, even on weekends."

In 1999, Thiner was asked by Al Kurtenbach and Gary Gramm to establish an office in Ankeny, Iowa, where one of Daktronics' biggest competitors Translux/FairPlay, was located. Thiner remembers that when he opened the Ankeny regional office in September of 1999, "It was a challenge being located only five miles from a major competitor's factory, but our orders for Iowa grew from $50,000 to over $500,000 in just three years." The success of his office is a result of his good work. He'd also been properly indoctrinated by his professors at South Dakota State University, the training staff at Daktronics, his experiences with Jones in the Seattle office, and the lessons he'd learned in Indianapolis.

Also in the late 1990s, Daktronics began a new and innovative program in Michigan involving a partnership of sorts with retired high school athletic directors who would assist regional offices in contacting high schools regarding new scoreboard purchases. To Al and Frank Kurtenbach, who had coached in Iowa and had contacts with athletic officials throughout the Midwest, Iowa was fertile ground for the development of a program in that state. Thiner, they believed, was a perfect fit to work with these former athletic directors.

Darrell Thiner, Jeff Gullisckson and John Paloma

Little did Thiner realize at the time that by working in concert with retired athletic directors in Iowa, a batting average of nearly 1,000 percent in scoreboard sales would be possible. Employing this group of respected and well-known individuals, first in Michigan and then elsewhere, was a first in the industry. So successful has the program proven to be that it is emulated by many competitors.

A slice of the idea of involving sports administrators who had spent a lifetime working in coaching and in guiding and governing high school sports had entered into Frank Kurtenbach's mind in the early 1990s as Thiner was just beginning to grow the Indianapolis market.

A former high school wresting coach and now Daktronics' Vice President of Sales, Kurtenbach was on a plane flying to Texas to attend an annual meeting of the National Intercollegiate Athletic Association. On that flight he happened to meet Art Newcomer, then the respected veteran athletic director of the Overland Park (Kansas) high school. During a conversation with him, Kurtenbach inquired if the national association of which Newcomer was then an officer could use a stipend from Daktronics to help in honoring regional winners of the association's then five Athletic Director of the Year awards. Newcomer liked the idea, and a plan was developed. This program not only was offered by Daktronics in the spirit of helping honor deserving individuals at the high school sports level, but it gave Daktronics much-needed visibility nationally within this important group. A form of this early stipend idea continues today, incidentally, although altered and adjusted to meet today's needs.

Heidi and Jeff Gullisckson at the Kurtenbach farm.

Later, Jeff Gullickson, Daktronics' office manager in Niles, Michigan, happened to inquire of Gramm and Frank Kurtenbach if he might employ the soon-to-be-retired athletic director at Auburn (Indiana) High School to assist him in contacting high schools in his sales and service area. Thus, on June 25, 2001, Dick McKean of Auburn, Indiana, became the very first of what would become the forty five-member group of retired athletic directors employed by Gramm in the High School Parks and Recreation Department. McKean remained with Daktronics until the summer of 2008 when he reduced his workload to part-time. These retired athletic directors visit with colleagues about scoreboard needs and if a sale of a scoreboard or other Daktronics product is possible, technical support from Daktronics office managers is the follow up. The program has been a resounding success.

Interestingly, among the membership today is the man Frank Kurtenbach met years earlier on the airplane flight, Art Newcomer. He is now retired and assisting in making contacts with high school athletic directors. Another retired athletic director helping out is Charles Turner, who was athletic director at Cedar Shoals High School in Athens, Georgia, for fourteen years. "Daktronics is the place to work after retirement," he said. "It is a low pressure job that allows me to work from home, but still do a lot of traveling." Les Wright, who retired after twenty-six years as athletic director at Floyd Central High School in New Albany, Indiana, says his work with Daktronics keeps him in touch with his peers. "After I retired I missed

Retired Athletic Directors (2007). Front row (left to right): Tom Giorgio (NJ), Butch Newman (MS), Gene Robertson (IN), Bruce Taylor (WA), Art Newcomer (KS), Steve Smiley (OK), Kirk Crochet (LA), Bill Newcomb (TX), Terry McCombs (MI), Bob Gershman (MI), Dick McKean (IN), Ken Green (MO), and Lynn Freshoul (OR). Middle row: Les Wright (IN), Dennis Coffel (MD), Jeff Paris (FL), Pete Vela (TX), Al Mallanda (NY), Harold Stephens (TX), Ron Gadus (AZ), Steve Montgomery (TX), Jim Hess (TX), Joe Richer (WA), Al Strand (WA), Bob Holsclaw (ID), John Huizenga (MI), and Frank Kurtenbach. Back row: Jim Loughran (CO), John Ford (NY), Dick Schumacher (KS), Dave Vogelgesang (IA), Charles Turner (GA), Dave Collison (IA), Jim Webb (AR), Don Poe (TX), Kelly Reeves (TX), Tim Moore (IL), Stan Gida (OR), Steve Kinney (KY), Hub Reed (OH), Barry Parsons (OH), and Gary Gramm. Not pictured: Bob Smith (CO), George Spencer (GA), and Dale Evans (SC).

the people I had worked with as an athletic director. Daktronics gave me the opportunity to still be involved in athletics," he said.

These athletic directors, from as far away as Secaucus, New Jersey, and Penbroke Pines, Florida, are invited to Brookings periodically to learn about new products and to discuss common challenges with Daktronics officials. It's also a social gathering where plenty of athletic shop talk and stories are also recounted.

Meanwhile, Thiner's successes in the Ankeny office can be partly attributable to the help he received from a retired athletic director. George Long, athletic director at Urbandale High School for thirty-nine years, was not an employee of Daktronics, but was a self-employed independent dealer, owner of George Long Company and a Daktronics dealer since 1988.

Long and his company's five employees worked closely with Thiner in contacts for Daktronics scoreboard sales throughout central Iowa. "I was receiving instant trust and credibility by being associated with George and he was getting the technical expertise he needed," Thiner said. By 2006, Long was ready to retire, and Daktronics bought his company.

Thiner said one of the most important aspects of making his office a success has been the relationship the retired athletic directors have with their friends still in the high school athletic fraternity. "They bring a trust factor that is not given to other salespersons," Thiner said. Helping Thiner with Iowa contacts now are retired Iowa Falls High School Athletic Director Dave Collison and Dave Vogelgesang, retired Athletic Director from Tipton High School.

Darrell Thiner

Thiner said Daktronics orders have grown in recent years with the advent of video boards at the high school level. Orders in Iowa have grown from $48,000 to over $1 million in Thiner's first eight years in the state. That's in a state with slightly fewer than three million people, and it is the home of one of Daktronics' strongest competitors.

By the mid-1990s, there were ten Daktronics scoreboard sales and service offices around the country, but it was becoming clear that Kurtenbach's original plan to establish four new offices a year needed to be adjusted. He had originally envisioned that each new office manager, who would be a graduate of South Dakota State University or a shining star from the Daktronics work force in Brookings, would spend time in customer service at the plant, then time in Seattle with Marlo Jones before branching out on their own. But this was time consuming. To ramp up the efforts at regional sales and service offices, a person might be hired first and then brought later to the plant for orientation.

But Bryan Nagel, Perry Grave and Paul Wildeman would still be trained and prepared the old-fashioned way for their adventures as pioneer sales and service representatives out on their own.

5

THE WHITE HOUSE CALLING

Growing up on a dairy farm near Gettysburg, South Dakota, named after the famous Civil War battle and the 272-word presidential statement about it, Bryan Nagel never dreamed that in 1998 he'd be coordinating a visual display to illustrate what the President of the United States was telling the world about the country's financial condition.

But there he was, on May 26, 1998, out of sight in an anteroom of the Executive Office building's press room, huddled about ten feet from the President's podium with fellow Daktronics employee Paul Kurtenbach. The two were operating the company's standard scoreboard that was connected to the Daktronics' Venus 1500 software, controlling the speed of a dollar-by-dollar message display countdown. The display was coordinated with President Bill Clinton's remarks, along with a few other top officials, as they told millions of viewers in this nation and beyond how the administration had turned a $357 billion deficit into a $39 billion surplus. The scoreboards' numbers were to be synchronized to reach $39 billion at exactly the moment President Clinton mentioned that figure.

After the announcement was over, the President stopped by to thank the two Daktronics employees for their twenty-one-hour ordeal spent preparing the visual part of the President's report.

Baltimore, Maryland, Daktronics Scoreboard Sales and Service van.

That was a highlight for Bryan Nagel, but he has many more fond memories of his work establishing a Daktronics Sales and Service office in Baltimore, Maryland, in late 1992. His career with Daktronics started in 1991 while in his final year at South Dakota State University majoring in Electronics Engineering Technology. He applied for and was hired by Daktronics as a customer service technician and spent the summer of 1991 in subassembly work on the assembly line.

He graduated in June of 1992 and moved into the customer service department to work in the repair center. Later, he was on the road for a time helping with installation and service of Daktronics products. He heard Daktronics CEO Al Kurtenbach discuss the opportunities available as a manager of a new Scoreboard Sales and Service office. "I knew it was a perfect fit for me," Nagel remembered. He had minored in economics with an emphasis on sales, and his engineering degree certainly gave him the "service" qualifications. "I was looking for a job that involved sales and service," he said.

Bryan Nagel

He applied for the position in August of 1992. He and his wife, Teresa, who grew up in Brookings but graduated from high school in Watertown, South Dakota, and their six-week-old son, rode a rented truck to Seattle for orientation and on-the-job training with Marlo Jones in preparation for Nagel's assignment establishing a new office in Baltimore. The next six months in the Northwest were spent in scoreboard sales and service work and in adjusting to life in the big city and its plethora of fast lanes, especially compared to the Nagel dairy farm near Gettysburg.

"It was a great experience," he recalls, "especially working with high schools all the way up to large venues such as the Seattle Center's Key Arena, home of the SuperSonics. I still remember making the monthly trips to a mall in Vancouver, Washington, to change light bulbs. Jones, his Seattle mentor, said Nagel was a perfect candidate for what he'd be doing in Bal-

Key Arena, Seattle Washington (1995).

timore. "I'm sure working in the Seattle office helped, but I am also quite sure that he learned much more about how to sell when he started growing the Baltimore office," Jones said. Nagel appreciated the patience, help, and guidance he received from Jones. "I'm thankful for Marlo's great leadership, as well as similar efforts by other office managers who, over the years, have helped me along the way," he said.

In December of 1992, Nagel and HSPR Market Manager Gary Gramm headed for Baltimore to look the place over and make preliminary arrangements for the fourth Daktronics Scoreboard Sales and Service office. "Talk about a daunting task," Nagel recalled. "Trying to get to know the area in a few days, finding a place for the office and for our new home was difficult." Gramm, as he would do many times during the next nearly two decades as new offices were launched everywhere, helped Nagel with the details. They found suitable office space that would be expanded into two larger suites in the same Linthicum, Maryland, business park a few years later. With the Baltimore stage set, Gramm and Nagel returned to Brookings where Nagel and his wife packed for the trip. There was also plenty of paperwork in preparation for the move, and Nagel also selected an inventory for his new office.

While in Brookings preparing for the move, word came that the famous university for the deaf and hard of hearing, Gallaudet, of Washington, D. C., had placed an order for a scoreboard and message board, and these were added to his inventory. The Gallaudet installation was Nagel's first out of his new office. Later, installing the equipment at Gallaudet, he met a helpful Kris Gould, the university's equipment manager, and the two remain close friends today.

Bryan Nagel, Rookie of the Year 1993-1994.

Finally, the young Nagel family was on its way, again in a rented truck, leaving the sub-zero Dakota February temperatures behind for "The Free State" with the warmer climate.

One of the factors in the selection of Baltimore as the fourth Daktronics satellite office was the famous Camden Yards, home of Daktronics' good customer, the Baltimore Orioles. The Orioles were the first major league baseball customer when it purchased a Daktronics scoreboard, an out-of-town scoreboard and a ribbon display board before Nagel arrived. Those in-place displays were invaluable in helping the new office gain visibility and make contacts. Nagel said that even though Daktronics had those tremendous displays at Camden Yards, most potential customers in the area had never heard of Daktronics. Nagel used the Camden Yards installations as stepping stones. "It was always helpful to tell customers that we had installed the state-of-the-art system at the new baseball stadium in town," Nagel said. Ironically, twelve years later when Nagel moved to set up an office in St. Louis, Missouri, he was able to employ a similar assertion in calling attention to another Daktronics creation at that city's Busch Stadium.

Nagel said his first few months in Baltimore were spent on the road calling on school administrators and athletic directors in Maryland and Delaware to inform everyone he could meet about Daktronics' "service first" program. "It was a great way to open doors and build relationships with athletic directors, and I started out with servicing relationships with

Baltimore Sales and Service van. Maryland Daktronics Sales and Service inventory.

a large number of them. Those relationships grew to scoreboard sales," he said. Nagel remembers that in the early years, Coca Cola® and Pepsi® were highly visible in the schools as sponsors of scoreboards. "I devoted many hours cultivating the relationship with both bottlers, and as time went by I can recall that bottler sales made up seventy-five percent of all scoreboard sales," he said.

When Nagel arrived in Baltimore, his long-term goal for his office was to stimulate more sales in the high school and parks and recreation markets than that of any other office. "I was able to reach that goal, and also became the first HSPR office to surpass $1 million in sales," he said proudly. "It was fulfilling for me, and I would challenge all office managers to set personal goals of this nature and then work hard to achieve them," he said.

It wasn't long before Daktronics officials recognized Nagel's good work. To speed up the process of starting new offices at the four per year pace, Daktronics started sending potential office managers to Baltimore for training. Nagel remembers working with future office managers Matt Heinse, who later opened an office in Baton Rouge, Louisiana, Paul Farley, who went on to Albany, New York, and Paul Kurtenbach, who succeeded Nagel when he went on to the office in St. Louis. "We all worked on some interesting installations," Nagel said, "including for the Washington Redskins, the Orioles, the U.S. Naval Academy, the University of Maryland, and many others."

His first installation out of the Baltimore office, after the Gallaudet work, was at Penn State at State College, Pennsylvania, which was a three-hour drive from his office. "I learned to make that drive in my sleep after about twenty treks in the first year," he said. At Penn State he installed new baseball and softball scoreboards, including the scoreboard at the Nittany Lions' Beaver Stadium, which is the second largest stadium in the United States and was one of Daktronics' first sports applications of Red, Green, Blue and White (RGBW) colored lamp displays. He remembered that project mostly for what went wrong, not right. "It didn't help when the electricians that wired up one of the boards failed to connect neutral to ground at the step-down transformer and destroyed all of the drivers in the display. I spent seven or eight days climbing up forty feet of scaffolding and removing drivers and then taking them back to a makeshift work bench on site to replace a few chips that were affected. This, along with changing hundreds of triacs on the main board, made for an exciting time in Pennsylvania," he said.

Matt Heinse training in the Baltimore office.

His association with the Redskins football team was also memorable. "I was able to help build some relationships with RFK Stadium by servicing its American Sign and Indicator equipment," he said. Nagel attended every Redskins football game from 1993 until the team moved to Jack Kent Cooke Stadium, later named FedEx Field. He said he'd like to forget the first year in that complex that had new Daktronics equipment. "I was at every game and they never won one at home."

In 1996, Nagel was on site for a major tennis tournament in Bethesda, Maryland, where new technology by Daktronics was in use in the form of the Magnaview display. "I was supposed to be on site for one day to make sure the stats interface worked out," he recalled. But it rained for two days and work was impossible. The upside for Nagel was spending time in the club rooms with "some very influential folks" while he waited for the weather to clear. He recalls sitting with tennis great Jimmy Connors, watching television with him. "He had some interesting stories about basketball star Michael Jordan's golf game. He was just a regular guy," Nagel said.

The U.S. Naval Academy has been a good friend of Daktronics since the mid-1990s. Nagel was involved in replacing two football scoreboard displays at the academy, as well as many other scoreboards for various sports. "All the people there were a pleasure to work with," he said.

Baltimore office work bench.

Nagel's most unusual experience was in May of 1998 at the nation's White House. "I remember sitting in a high school somewhere and at about 1 p.m., getting a phone call from Daktronics in Brookings," he said. Nagel was told the White House had called Daktronics to rent a display for a press conference scheduled for 10 a.m. the next day. Nagel was to handle the job. Paul Kurtenbach, a full-time Daktronics technician who was also working on a technical degree at Capitol College at Laurel, Maryland, worked with him. Together they worked their way through an exciting and pressure-packed twenty-one hours. That night, Nagel and Kurtenbach (unrelated to company co-founder Al Kurtenbach) picked up the necessary equipment that had been flown out from Brookings and drove to the Executive Office building in Washington.

It was about 10 p.m. when they met representatives of the President's press staff, who explained in more detail what they wanted. Nagel and Kurtenbach learned the President would have a press conference to announce that what had been a $357 billion budget deficit when he took office had become a $39 billion surplus, or would reach that figure in twenty-one-hours. The Daktronics display would count down the dollars as the president and others spoke, and would display $39 billion when the President arrived at that part of his speech where he mentioned that surplus amount.

Bryan Nagel and Paul Kurtenbach spent 36 hours straight getting the sign ready for President Clinton's speech.

Nagel and Kurtenbach took the display back to the Baltimore office to "get things figured out," Nagel remembered. They worked most of the night and made several post-midnight phone calls to technicians in Brookings to check on some details. Helping from Brookings were Ron Koerner, now head design engineer for Galaxy® Display Products, and Jodi Koerner, region sales coordinator for the commercial market who had worked with Nagel in customer service.

By 7 a.m. that morning of the press conference they were back at the Capitol ready to go. The two men, their toolboxes, and the Daktronics equipment all survived polite but thorough searches, including an x-ray scan of the equipment by Capitol security personnel. "It was like your basic airport search," Nagel remembered. By 9 a.m., Nagel and Kurtenbach had everything ready. "We then sat in a room about fifteen feet from where the speakers were. As the speeches progressed we were told to adjust the speed of the countdown to be synchronized with certain points in the speeches," Nagel remembered. The AS-4000 was programmed to change the counting speed, with a start value of zero and a count value of 10,000 per second, and then the speed was adjusted accordingly as the President's staff directed, either slowing up the message display, or speeding it up to match what was being said to the press corps.

Bryan Nagel shaking President Clinton's hand after his speech.

It all went off without a hitch, and was an exciting event for Nagel and Kurtenbach. It was broadcast on about sixty television channels. Recognizing the efforts of Daktronics in Brookings, and those of Nagel and Kurtenbach on the site, White House staffers agreed that the Daktronics logo could remain on the equipment in the pressroom that was being photographed and filmed. After the conference, President Clinton agreed to pose for a picture with the two Daktronics representatives, and to take a moment to thank them for their all-night efforts. "I honestly don't remember what he said to us," Nagel commented, "but it was pretty short...maybe something like 'nice job,' I guess."

In August of 2004, Nagel left to assume similar duties and responsibilities in St. Louis, and Kurtenbach assumed responsibilities for the Baltimore office; a position he still holds in addition to being Daktronics' New England/Mid-Atlantic region manager for HSPR.

6 PYROTECHNICS

Perry Grave grew up in Brookings and graduated from Brookings High School in 1987. He hoped to become an architect. But after high school he enrolled in the South Dakota State University College of Engineering and was well on his way to earning a degree in Electrical Engineering. During his junior year, he considered changing his electrical engineering major to one more conducive to the eventual earning of a degree in architecture. But his advisor, Dr. Virgil Ellerbruch, talked him out of it. "It was good advice and I'm happy with my career," Graves said. As a student he worked part-time at Daktronics in manufacturing and customer service departments, and then became a full-time employee after graduating in 1992. He decided to enter the Daktronics training program in what then was known as Scoreboard Sales and Service.

He was about halfway through the training program in Seattle, Washington, working with Marlo Jones, when he learned of an opening in Oklahoma. He made his interests in that post known and was selected by Daktronics to carry on the pioneering efforts of that early cadre of sales and service managers by setting up the first Daktronics Scoreboard Sales and Service office in the South Central Region.

"Perry did a good job with customer relationships in the Seattle area,"

Perry Grave in front of the Norman, Oklahoma office (1994).

Jones recalls, "and it wasn't too long before he was managing the new office in Norman."

Grave arrived in Norman, Oklahoma, in April of 1994. "We were a one-man show then for about two years," Grave said. He was picking up where Lisa Kurtenbach Glanzer had left off. Glanzer, the daughter of Daktronics co-founder Al Kurtenbach, is a 1990 graduate of South Dakota State University with a degree in mathematics.

Her husband was in the Air Force and was stationed in Norman. She sold Daktronics scoreboards part time from her home, calling on area high schools. Glanzer, incidentally, was a starting guard for the university's womens basketball team in the late 1980s and still holds records for three-point baskets made. Hers was not a full-fledged sales and service effort, but she did establish a Daktronics presence in the Norman, Oklahoma area.

Perry Grave working diligently at his desk.

In early 1994 shortly before her Air Force husband was scheduled to be transferred, she notified Directors of Sales back in Brookings that she would be leaving Oklahoma, considered by Gary Gramm, Daktronics HSPR market manager, as a prime area for the next Sales and Service effort. "I talked to Perry Grave, who had expressed an interest in our sales and service program there," Graham said. "I encouraged him to take the assignment at least on a temporary basis to see how it goes." Grave agreed and moved to Oklahoma to continue and expand upon Lisa Glanzer's ground breaking work. "As it worked out, I discovered I really liked Oklahoma and the people," Grave said. "After a few years I picked up the accent and got used to the hot summers and I have never really thought about moving anywhere else."

When he arrived in the Sooner State, with the help of Gramm, he rented a small, 1,500 square-foot office/work area and from it he serviced and sold in the Norman area, becoming well acquainted with many in the

University of Oklahoma Athletic Department. Over the years, his association with the University of Oklahoma has resulted in many projects at that prestigious university.

The market in Oklahoma improved annually for Daktronics, thanks to the efforts of Grave who, after about two years working solo, was joined by others who journeyed south to work with him in what would become the first office to reach $2 million in sales in fiscal year 2004. "We weren't the largest Daktronics Sales and Service office by any means, so we were very proud of that accomplishment," Grave said.

Grave also expressed pride in his part in providing one of the first Daktronics ProStar® systems in Oklahoma in 1997. The thirty-two foot wide by one hundred sixty-five foot wide scoreboard included a twenty by thirty-foot screen and was placed at the University of Oklahoma. Although the initial sales contacts for the ProStar® were made by others from Daktronics, Grave was included in the sales team later on and sat in at the various meetings with University of Oklahoma athletic offices. The fact that his office was close by and available for almost instant service should something go wrong with the scoreboard helped seal the deal. "It was nice to be a part of the launching, so to speak, of Daktronics into the next level with the ProStar®," he said. Since that 1997 sale, the University of Oklahoma has been a "very good customer for Daktronics," he said.

Grave said he has always approached customers with

Perry Grave at the work bench.

what has become the Daktronics trademark, that of caring and helpfulness. "The customer always comes first in my mind," he said. "The customer is why I have a job and why everyone at Daktronics here in Oklahoma or back in Brookings has a job." He said his customers are the reason Daktronics "is where it's at." And Grave doesn't limit the placing of importance only on

Storage area in Norman, Oklahoma.

the decision makers of the organization or company with whom he is working. "Everyone you encounter on the job, top to bottom, is important."

One of Grave's most embarrassing moments came early in his career in Oklahoma. He was asked to make a game/shot clock presentation to a group of high school athletic directors and coaches in the Frontier Conference in northeast Oklahoma. They were considering going to the use of a shot clock. Grave brought a Daktronics shot clock with him for his presentation. He set it up on a table and plugged it in so that it would be running as he told about its advantages and how it operated. He stepped in front of the table and started to address the audience, which he said was very attentive.

"Then I started to notice a curious look on some of the people in the first few rows, and they pointed to the shot clock behind me. When I looked back, smoke was pouring out of it," he said. He quickly unplugged it, and with a smile he told his audience that he always provided "pyrotechnics" to get people's attention during his presentations. The malfunction was due to resistor failure, and he was able to explain this and work himself through the embarrassment of the moment. Everyone in the audience appreciated his candor. The excitement apparently added to his presentation. "I got three sales out of it," he joked.

Grave is currently not only the office manager in Oklahoma, but is heavily involved in three markets—Commercial Schools and Theaters and Large Sport Venues. He also remains involved in some large high school projects in the area. He is assisted by a staff of seven, including two student interns. The original Daktronics Sales and Service office space has doubled since he arrived, and despite the smoking shot clock incident, his loyal customer base has grown each year.

Grave's geographic area of focus has also been enlarged to include Oklahoma and Arkansas. Among installations with which he's been closely associated are the systems at Tulsa Union High School, the University of Oklahoma, Oklahoma State University, the ASA Hall of Fame Stadium and Ford Center Arena, both in Oklahoma City, and Blackwell, Alrus and McAlcester High Schools, plus many others.

7

Wrestling, Wedding and WordPerfect

At six feet, four inches and over 250 pounds, Paul Wildeman of Cherokee, Iowa, spent hours as a student at South Dakota State University glancing up at the old, rectangular scoreboard at Coughlin-Alumni Stadium that was always there for him and his football teammates. The old and ordinary scoreboard, at least a quarter century into its life by then, has since been replaced with a large, colorful scoreboard and video display having most of the bells and whistles in the Daktronics arsenal. Now as manager of the San Antonio, Texas, Daktronics office and also responsible for the South Central Region and New Mexico, Wildeman has logged thousands more hours to his scoreboard watching record, but this time not as a star football player, but with the critical eye of a knowledgeable Daktronics official. He's seen them all, and more often than not, he has installed and repaired a good number of them, including many similar to the new system now lighting up the old field where he was a star lineman.

Paul Wildeman in the San Antonio office (1999).

He played football at South Dakota State from 1987 to 1991, earning all North Central Conference honors for his good efforts. He was a high school heavyweight champion wrestler for his hometown of Cherokee, Iowa, and he also continued in that sport, albeit a brief tenure, at South Dakota State University. Ironically, in his second week of collegiate wrestling practice, he was working with another heavyweight and suffered a wrestling career-ending injury, compliments of Steve Kurtenbach, the son of Frank Kurtenbach, who was the National Sales Manager for Daktronics. Frank Kurtenbach has worked closely at times with Wildeman, his son's opponent on that fateful day of practice.

Wildeman majored in mechanical engineering at South Dakota State University. After graduating in 1992 he joined Daktronics as a trainee for a future Sales and Service manager position. After spending time working in manufacturing, getting to know the company, its products, and its culture, he assisted in Customer Service and on the Daktronics Help Desk. He also did summer service rotations. He recalls that on the summer service trips, some took him to new territory, traveling the long, straight highways of North and South Dakota. "It was my first time traveling through the western parts of these states and I couldn't help but feel a sense of adventure out there," he remembered. It was on one of those trips that he acquired a strong sense of pride in representing Daktronics, "and I learned a lot from the experience."

San Antonio office (1999).

When Daktronics offered him the opportunity to move to San Antonio, Texas, to establish a new Scoreboard Sales and Service office there, he quickly accepted. He wasn't disappointed on what he called his "adventure." He said the people of the area were friendly and "it didn't take long for my wife and I to consider San Antonio home." Wildeman said the opportunity to represent Daktronics in the region "has been very gratifying." Although Daktronics was once a name that was seldom heard in Texas, that's changed. Because of Wildeman's good work, Daktronics is now one of the leading providers of scoreboards, video screens, and message centers in his region.

But before traveling to San Antonio and starting off on his own, in February of 1994, Wildeman left for Seattle to work with Marlo Jones. "That's where I really learned to run a regional office," he said. That year of 1994 was also an eventful one for him. Wildeman's future wife, Linda, graduated that May from South Dakota State University with a degree in Nursing. Linda then returned to her home in Colorado to plan for their wedding set for late May. Wildeman took a few days off from his Seattle work and returned for that happy day. Then the newlyweds returned to Seattle where he resumed his training with Jones. As a mechanical engineer he lacked some of the formal training as an electronic technician. "But he did a great job and learned all that was necessary to service scoreboards," remembered Jones.

San Antonio storage.

Wildeman's adventure was about to begin. "In August of 1994 Linda and I packed up what little we owned and drove to San Antonio," he said. As one of the early scoreboard sales and service managers, he and the other pioneers in the field and those in the Brookings office, such as Gramm, Weninger and Lynette Smith, were still learning. "Over time," he recalled, "procedures and policies were refined and have become more clearly defined." When he opened the San Antonio office, "We were just trying to learn from each other and do the best we could with the tools and the collective experiences we shared as a group," he said.

By today's standards, Wildeman's early years were in the technological dark ages. His first office computer had an eighty megabyte hard drive and WordPerfect® software on which he did his quotes. There were no templates, no DakPipe® software, "just a price sheet and a calculator with which to figure out discounts," he said. "I just hoped that I could remember to include all the accessories, shipping costs, installation and the rest so I didn't make a costly and embarrassing error."

While his upper Midwest accent never resulted in misunderstandings with customers, it often slowed the conversation down, but usually to everyone's enjoyment. "When the movie *Fargo* came out, I was the hit of my social circle for my ability to imitate the movie characters," Wildeman said. Once, when Wildeman and Linda were house hunting, the sales agent asked what Wildeman did for a living. He told her and she then asked if he used any "lie-zahs" in his work. Not until his wife nudged him in the ribs and explained the question did he realize the woman was asking about "lasers."

To help him in his managerial responsibilities at his office, he earned an MBA degree in 2003. "I kept chipping away at it in the evenings and some weekends until I finished," he said. Along the way, Daktronics was supportive with encouragement and financial help, for which he is grateful.

During his time in San Antonio, Wildeman has been involved in many projects, including the Texas House of Representatives voting system, scoreboards and message boards at Southwest Texas University, and literally hundreds of displays in high schools throughout the region plus many other venues.

One of his favorite projects was the Honey Bowl football project at Uvalde, Texas. He remembered it not so much because it was technically a difficult engineering challenge, but because during the selection process it pitted Daktronics against a well-known competitor in the Texas area. "We badly beat a small regional competitor company, and they were fuming mad," he said. "They made accusations to the school district and called us a 'fly-by-night' company. Needless to say, their tactics didn't work and we gained a repeat customer."

Wildeman and his compatriots in the late 1980s and early 1990s worked through the newly-established Scoreboard Sales and Service model to help move Daktronics into the national market. With the growth and success of Daktronics, no one today would even consider referring to the company as a "fly-by-night" operation. This handful of young, eager Daktronics disciples would pave the way for the impressive network of Daktronics Sales and Services offices of today. They helped craft and carry forward the Daktronics tradition of helpfulness.

It wasn't easy for them, but to a person they say they left Brookings determined to work the long, hard hours and make a success of their efforts. By 1994, Daktronics had ten regional offices successfully up and running.

Those veterans and the hundreds more who have followed in the establishment of new offices in other parts of the nation have represented Daktronics in impressive fashion. Few companies anywhere enjoy the reputation for helpfulness and good citizenship that the Daktronics family has established and maintains. What follows are the stories of some of the experiences of many others who proudly carry on the Daktronics tradition of quality, caring, and helpfulness.

Going The Extra Mile

They arrive daily at the Daktronics plant in Brookings, or at a regional office somewhere in the USA. Words of praise and commendation for Daktronics projects large and small are always appreciated, and Daktronics products, service and the people of Daktronics seem to inspire customers to sit down and write a letter. The equipment is excellent, everyone writes. It was installed on time and on budget, as promised. And invariably, there is heaping praise for the people of Daktronics, from the designers, builders, installers and those who service. One customer even wrote to commend the workers who had helped him load a small scoreboard onto his trailer.

Everyone at Daktronics, it seems, goes the extra mile, from the CEO to the dock workers who helped a booster club president from Iowa load a scoreboard onto his trailer for the long haul back to his hometown high school track.

Following are examples of actual letters Daktronics and its region sales and service offices receive every day. Space did not permit printing excerpts from every one, but the following samples give some idea of the adherence of everyone at Daktronics, every day on every project, to provide extra special service no matter the challenge.

In 2004, Daktronics installed the first "ProRail" facia in the Glendale Arena in Arizona, as well as a center-hung scoreboard and an outdoor marquee.

"The system Daktronics designed and manufactured for us is a perfect fit for the arena. We are very pleased with the selection of Daktronics and with the final results of all of our electronic displays." Anthony P. Cosentino, vice president, The Ellman Companies.

Scott Carnahan, athletic director of Linfield College, McMinnville, Oregon, took the time in October of 2004 to write to Frank Kurtenbach, vice president for sales:

"I just want you to know how extremely happy we are with our new scoreboard and Pro-Star Video. We obviously are still on a very steep learning curve, but we couldn't be happier with all phases of the product, the sales and the service. Steve Davis, Nate Fossell and your staff in Brookings made this project come off without a hitch from the compressed delivery schedule, through installation and the on-site training."

Jeffrey Steele, associate athletic director at Auburn University, Alabama, wrote in late 2007:

"Daktronics has provided us with the highest quality, state of the art products and support for all our scoring, timing and information systems while giving us the ability to interface all technology into an exciting and revealing game-day production."

Roy Sutherland, director of the Mayo Civic Center in Rochester, Minnesota, wrote in August of 1988:

"The Rochester Park and Recreation Department and the Mayo Civic Center are thrilled with the equipment Daktronics supplied to the City of

Rochester. This equipment is 'state of the art' and has more than met our expectations...Your operator schools were very well received by our staff. The only limitation on the message boards is one's imagination."

Northwest District leader Marlo Jones received this letter from Kirc Roland, athletic director of Lower Columbia College, Longview, Washington.

"We are thrilled with the new basketball scoreboards and the new baseball scoreboard looks great. I want you to know how much we appreciate Steve Davis and Bruce Taylor. Bruce made the first contact here when we didn't have funding yet. He was patient, helpful and available. Bruce knows how athletic departments work, and he was fun to work with. Steve really is a pro. He not only helped me with the project, but he got the campus service crew everything they needed to make the installation go smoothly. When Steve came to the scoreboard christening, it showed me how much Daktronics cares about our school and our scoreboard project."

Matthew Jordan of Jordan and McCallum Insurance with offices in twelve southern communities headquartering in Anderson, South Carolina, wrote to tell Daktronics:

"Even our major competitors are contacting us to comment on the effectiveness of the Daktronics sign. Other businesses ask how they can get one for themselves. I give them the phone number and local distributor's name and tell them about all the options and exceptional services Daktronics offers."

Dean M. Ekeren, aquatics director at Texas A&M, dropped Al Kurtenbach of Daktronics this note in 2000:

"I just wanted to thank you once again for the outstanding products and services your company has provided over the years to our aquatic timing and scoring systems. Congratulations on a great staff dedicated to producing the finest equipment."

David Haas, general manager of the Hickory, North Carolina Crawdads, a Class A Affiliate with the Pittsburgh Pirates, wrote in 2004:

"Just wanted to let you know how pleased we were with the Daktronics team. Everyone we came in contact with was professional, courteous and friendly."

Bill Rowe, Jr., director of athletics at Southwest Missouri State University in Springfield, Missouri, in August of 1991 wrote a letter to colleagues to introduce Dave Marsh, director of the Scoreboard Sales and Service office.

"We at Southwest Missouri State University have been familiar with Dave since 1984 when he served as the installation supervisor for Daktronics when they installed new scoreboards in Briggs Stadium and Hammons Student Center, which serves the needs of our basketball program...We have on rare occasions needed to call Dave on short notice for emergency purposes and to date, we have been very fortunate as we have been able to get in touch with him and he has come to service our problem just as soon as he could."

Al Kurtenbach of Daktronics received this letter in 1995 from Jay James, account manager for Giltspur:

"More often than not during the course of business we are quick to find fault by negligent when it comes to giving credit where credit is due. Recently I had the opportunity to work on a project with one of my clients that employed the use of a Daktronics LED sign. My company, Giltspur, designs, builds and services trade show exhibits. The client wanted to *rent* an LED sign for a single trade show. In my opinion, the project was annoying in terms of the fact that it required a lot of work for very little monetary reward. Daktronics was able to accommodate and surpass my needs with hardware, software and programming services. Kelly Steinlicht in the "animation department" impressed me with her unwavering good attitude, "can do" spirit, and customer orientation. I wish my team at Giltspur had ten champions like her."

Wrote Michael A. Bobinski, director of athletics at Xavier University, Cincinnati, Ohio:

"Our experience with Daktronics has been overwhelmingly positive."

HSPR Market Manager Gary Gramm received this letter in the fall of 2004 from the College of Saint Rose in Albany, New York:

"The customer service I received from Paul Farley and Jenny Erickson was outstanding and I wanted to be sure you are aware of it. Many thanks to Paul, Jenny and all those Daktronics employees behind the scenes that not only manufactured our scoreboards, but then ensured prompt delivery."

Rocky Mountain Region leader Mark Johnson heard from Bob DeCarolis, athletic director of Oregon State University, shortly after installing the school's new integrated scoring and video display system:

"I would like to take this opportunity to express our sincere appreciation for the exemplary performance on the design, implementation and engineering and installation... Please do not hesitate to refer any of your future clients to me to discuss the outstanding quality of the Daktronics products, employees and representatives."

Bob Bierscheid, representing the Roseville, Minnesota, Department of Parks and Recreation, wrote Daktronics regarding what he called "our super scoreboard."

"We look forward to a long and positive relationship with Daktronics. We'd appreciate very much the schedule for upcoming training sessions and a catalog of package programs. Once again, please express our appreciation to members of your staff for all of their great help."

Meredith Camp, director of Events and Special Projects for the Cleveland Browns football team wrote to Daktronics President Jim Morgan in 2002:

"Daktronics has been extremely responsive to our needs and any problems that have arisen throughout the three years in the Cleveland Browns Stadium. You have always made every effort to troubleshoot and address issues in a timely manner."

This letter of praise was written by Sean McDonald, owner of Generations Restaurant in Avon, Massachusetts:

Just before the July 4, 2000, weekend I installed a Daktronics Galaxy® outdoor message center at my restaurant in Avon, Massachusetts. Anticipating a slow July, I was surprised when business suddenly increased. I attribute this boost and continual increase in business to the Daktronics message center, which is visible from the main highway."

In late 2002, Daktronics' Ginny Hereid, Timing and Marketing, received a letter from G. Robert Bettner, executive director of the College Swimming Coaches Association of American, Inc., in Colorado Springs, Colorado. Included in the letter of praise was this:

"It is companies like Daktronics, through its sponsorships that make it possible for the CSCAA to operate and provide services to all collegiate swimming coaches. The CSCAA looks forward to a continuing relationship with Daktronics in the future. The relationship has been beneficial to both organizations and hopefully will continue to grow and prosper in the future."

Allen C. Vella, general manager of Saginaw County Event Center in Saginaw, Michigan, wrote to Will Ellerbruch in 2002:

"In the last six months with the help of Michael Godoy at SMG, I've helped to purchase and or secure over $3,000,000 worth of products and services. Daktronics is one of the few vendors that has shown significant interest in our project after their product has been installed. I look forward to working with Daktronics far into the future."

Pocahontas, Iowa, attorney Dennis Fitzgerald, who was president of the Pocahontas Area Athletic Booster Club that purchased a Daktronics Omni Sport track timing board in the spring of 1994, wrote a three-page letter of praise to Daktronics President Al Kurtenbach.

"The purpose of this letter is simply to advise you that I have been thoroughly impressed with your entire organization. Both Kelly Koenig and Scott Dieck have been exemplary to deal with. Scott Jacobson was both knowledgeable and helpful in every respect. Paul Vugteveen was extremely cordial, friendly and clearly an asset to your organization. Even the gentlemen who assisted us in loading (the timer) into the trailer were quite courteous."

The Armory Foundation in New York City had a Daktronics scoreboard installed in late 2007 within the New Balance Track and Field Center in the Armory. CEO Norbert W. Sander, Jr., M.D. was pleased, and wrote to Daktronics CEO Jim Morgan:

"I would particularly like to commend you on the excellent staff you have at Daktronics who have sheparded this project through from its conception stage three years ago to the present completion. Our fans truly love it."

Michael Fox, sports director of the Indiana Hoosier Dome, wrote in 1991:

"The Hoosier Dome and the City of Indianapolis are being billed as the best site for a Final Four, due to team spirit and cooperation. Groups such as yours can hold their heads high, as you were a large part of this success."

James M. Downs, director of logistics for the U.S. Olympic Festival-1994 in St. Louis, Missouri, wrote in July of 1994.

"You made it happen! Without your support, the U.S. Olympic Festival-1994 could not have been the huge success that it was...I certainly hope we will have the opportunity to do business with your again."

Michael Thurman, College Center Operations Manager of Salt Lake Community College, took the time to write about his new campus sign installed in the summer of 1991.

"Our sign is a wonderful addition to the campus of Salt Lake Community College. Please accept our appreciation for a job well done in all areas."

Steven C. Maki, director of facilities and engineering at the Hubert H. Humphrey Metrodome in Minneapolis wrote to Daktronics CEO Jim Morgan in 2004:

"While the equipment your firm has provided has been absolutely fantastic, the service we receive from your employees has been equally impressive."

Ed Poppell, associate vice-president for Administrative Affairs at the University of Florida in Gainesville, in 1995, wrote Daktronics:

"As I write this letter, the University of Florida has just completed hosting the annual Gator Bowl football game...The Daktronics board was viewed by an extra 150,000 people this year over last year...Daktronics has made it easy and fun and I really appreciate you and your animation staff for giving our sequences that extra personality and flare."

Randy Doolittle, corporate director of purchasing for Skagit Valley Casino Resort in Bow, Washington, wrote Daktronics in November of 2003.

"From sale to installation, Daktronics offered assistance that exceeded our expec-

tations...After trying a different company, we came back to Daktronics and found the best product for our advertising needs. Daktronics stands behind its work, has assurance in quality, expertise and an exceptional maintenance guarantee."

George P. Foster, president of the Tacoma Tigers Baseball Club in Tacoma, Washington, wrote to Daktronics President Al Kurtenbach in March of 1994.

"I am writing to express my thanks to you and your employees for the week of training on our new scoreboard. While we were in Brookings we were shown a great time by all of your employees. They went out of their way to show us a great time in the evening when we had nothing to do. You need to be proud of them for showing us what customer satisfaction means. They went beyond the call of duty."

Michael J. Delano, owner representative of Comerica Park, wrote to Daktronics President Jim Morgan in 2002:

"Your project team hit a home run for us. They did an excellent job of installing a sophisticated scoring and matrix display system on time and on budget for the Tigers and their fans."

Mark Dreibelbis, assistant athletic director at Appalachian State University in Boone, North Carolina, took the time to write in 2000:

"The technical service provided by Daktronics is superb. It is the reason we purchased the new baseball scoreboard and purchased four scoreboards for the convocation center. We are committed to Daktronics through your service and excellent product."

In September of 2004, Bob Fischer paused from this duties as president of Fryn' Pan Family Restaurants in Sioux Falls, to write a letter of praise.

"In December 2003 we installed a marquee electronic sign from Daktronics at our Fryn' Pan Family Restaurant in Sioux Falls, South Dakota. The sign has caused more comments from the public than anything else we've done to the property, including remodeling. The multiple text fonts and sizes that are available help make the sign exciting, but the slide transitions and scrolling capabilities really draw attention."

Along Interstate 90, over Snoqualmie Pass in Washington, thirteen Daktronics walk-in full matrix LED Variable Message Signs were in operation in 1996 when fifty feet of accumulated snow caused traffic headaches and power outages. Wrote Steve Breyfogle of the Washington Department of Transportation:

"Other than local power outages, the Daktronics signs were consistently on line. In short, the signs have been very good performers during their two years of service." He also praised Daktronics service technicians for on-site help.

Gary Wyant, senior associate director of athletics at the University of Tennessee in Knoxville, wrote words of praise for their new Daktronics swimming scoreboard:

"The entire process from purchase to installation ran very smoothly. Again, we are most appreciative of the entire staff and of the support of Daktronics during the entire process. Thanks again from all of us here at the University of Tennessee to you and the entire Daktronics staff."

Scott Andraschko, sign and lighting director at Grand Casino, was impressed with all that is Daktronics, as expressed in this September of 2003 letter.

"We have purchased ten Daktronics displays throughout the years, including four ProStar® VideoPlus LED displays. We are thoroughly impressed with the service and support we have received from Daktronics and their employees. Reputation, presentation, reliability, and quality are only some of the reasons why we have ensured our loyalty in your company…I especially commend Brian Tucker, Brent Jofer and Keri Weinacht for their superior efforts and support to Grand Casino and myself. Daktronics really cares about product and quality in and out of the office—your staff is like family to me."

Michael Plant, executive vice president of business operations for the Atlanta Braves, took time out from his busy schedule in 2007 to write to Daktronics President and CEO Jim Morgan:

"Jim, the actions of your team and unwavering commitment to provide us with consistent performance in our equipment was never in doubt. I'm pleased to report that the boards have worked exceptionally well during this entire season."

Wrote Rob Harman, facilities supervisor for the City of Phoenix, about the scoreboard and video matrix at Phoenix Municipal Stadium in 1999:

"It is with no reservation at all that I recommend Daktronics as the scoreboard of choice with the City of Phoenix. I want to thank you (Rusty Lenners), Gregg Selberg and Jason Moen for your help throughout the process."

Pat Hagele, recreation services administrator for the Santa Clara Sports Complex-Aquatic Center, wrote to Daktronics in late 2004.

"Throughout the past year we have had excellent support in all ways from Daktronics, especially from the company's local office, and from Mike Gibbons in particular. With our Daktronics Timing System, along with Mike's dedication and commitment to providing exceptional service, the Aquatic Center has been able to effectively host the type of event it was designed for. A job well done."

Dan McShannock, athletic director at H.H. Dow High School in Midland, Michigan, wrote to Gary Gramm at Daktronics in 2006:

"I have been in interscholastic athletics for the past 34 years and in athletic administration for the past 23 years. During this time, I have worked with many sales representatives on a variety of purchases and issues, but your staff has ranked at the top for its service and professionalism."

Marlo Jones in Seattle received a glowing letter from Harold Phillips, Douglas County Fairgrounds manager, Roseburg, Oregon, in November of 2003.

"It has been three weeks now since our new Daktronics marquee has been installed at our facility. Not only are we pleased with the finished product, but so is the traveling public. As a public entity, I was careful about getting the right sign for the right price and this colored LED display was and is a great winner."

Dr. Richard McDuffie, director of athletics at Eastern Illinois University in Charleston, Illinois, wrote to Daktronics President James Morgan in September 2002.

"Eastern Illinois University, in partnership with DSN, now showcase one of the finest indoor video display systems in the region, due primarily to the professional determination and hard work of LeAnn Holler. Our fans and sponsors are extremely happy with the recent addition of new video, timing and scoreboard technology placed within our sports facility...Thank you again for all that you and your team have brought to the campus of Eastern Illinois University. We look forward to future endeavors with Daktronics and Daktronics Sports Marketing."

Randy Lewandowski, assistant general manager-operations for the Indianapolis Indians, wrote to Daktronics in 2003:

"Again, we thank you for the project team and for your efforts on the entire project. The cooperation of the sales, project management, service and installation teams could not have been better. We are pleased to have partnered with Daktronics again as we continue to keep Victory Field a first-class facility."

Dr. Charles A. Steinberg, senior vice president/public affairs for the San Diego Padres, commenting on the baseball software package the club starting using in 1998. It was based on a design at Oriole Park at Camden Yards.

"I am eager to introduce it to fans in San Diego. I've never regretted taking that chance on Daktronics long ago and trust it has been a rewarding and prosperous experience as you've become the dominant force in the industry."

Fred Safstrom, executive director of the Everett Public Facilities District in Everett, Washington, took the time in early 2004 to write to Daktronics President and CEO Jim Morgan.

"The bid award (for scoreboard, hoist, controls and peripherals) was made to Daktronics on May 28, 2003, or just four months before our scheduled opening of October 3. Daktronics proceeded to design and gain approval for an efficient hoist system and completed the scoreboard installation in the month of September. After opening, when we experienced some difficulty receiving the signal from the remote camera, Daktronics stepped up to help solve the problem by installation of an additional antenna. Again, the work was performed at no additional cost to the district...In a word, we are very pleased with Daktronics' system."

Jonathan Lee, marketing director for the Los Angeles Memorial Coliseum and Sports Arena, wrote in November of 2003:

"Where there was once an outdated, unattractive and unreliable incandescent structure that was turned on for events with fingers crossed, now stands a dynamic, eye-catching LED scoreboard, with capabilities far beyond what we had before. I am sure it surprised our patrons the first few times they saw the brightness, higher resolution and graphics presentations that we now enjoy...I just wanted to take the time to convey my satisfaction with just how well the LED scoreboard project turned out."

Jerry Pufall, associate athletic director of the University of Arkansas Razorbacks wrote to Robert Dedman of Daktronics in early 2004:

"On a personal note, I would like to thank you for your honesty and professionalism throughout the entire process. From the selection of your company to provide our scoreboards and video displays through the installation and follow-up services, you have represented yourself and your company with the highest integrity."

Ken VanSickle took the time to write to Marlo Jones in his Seattle office about the scoreboards installed at University High School, Spokane Valley, Washington.

"We have had few problems with your scoreboards. They have been very user-friendly and any problems we did have were promptly addressed by either Joe Richer or your tech...I would recommend Daktronics, Inc. to any school that is considering a scoreboard or reader board because of the quality of the product and the service after purchase."

Luis Carlos Estrada G., multiservice LCE, C.A., Barquisimeo, Venezuela, wrote in late 2001.

"I would like to send my word of appreciation to Daktronics as a whole and to the outstanding team of professionals led by Seth Hansen, Bryan Jacobson and Becky Misar, who made it possible to perform a series of installations in record time with very short notice...I knew well before the games that the quality backing up the Daktronics products was outstanding, but I knew very little about the world class, all star human and professional quality of your staff, you have a winning team, keep it that way."

Anthony Cillo, director of athletics at the University of Denver, wrote:

"Thank you for your special efforts seeing we were well taken care of relative to completing installation of the Daktronics scoreboard equipment package at the University of Denver."

This came in July of 1991, from Morry Stillwell of the United States Figure Skating Association in Malibu, California.

"This letter is to express the appreciation of my Association in the professional manner in which your staff has processed our recent order for a Marks Display System to be used during the USFSA-sponsored Figure Skating events."

Jack Martin is facilities director of the Alton School District in Alton, Illinois. He wrote to his fellow school administrators on behalf of Daktronics. Part of what he included in his letter was this:

"When we went with Daktronics I was immediately impressed. In my position, product and service are extremely important. Byran Nagel is the regional manager and works out of the Earth City office. Bryan is also the only local, full-time scoreboard and message center technician. I have found Bryan to be extremely knowledgeable and a tremendous source throughout our scoreboard project."

Neil Goter, athletic director at Wagner Community School District in Wagner, South Dakota, wrote to the Daktronics home office in September of 2007.

"We recently purchased new scoreboards as well as message boards for the gym and for an outside sign. I couldn't have been any more pleased with the service I received from the Daktronics people, especially Don Hansen."

Wrote Kurt Simon at Carmi-White County High School, Carmi, Illinois, in early 2008:

"Thank you, Daktronics, for all of your assistance in improving our football and basketball programs. We will enjoy the scoreboard, scorer's table and sound system, and we will remember your assistance and cooperation every time our scoreboards light up."

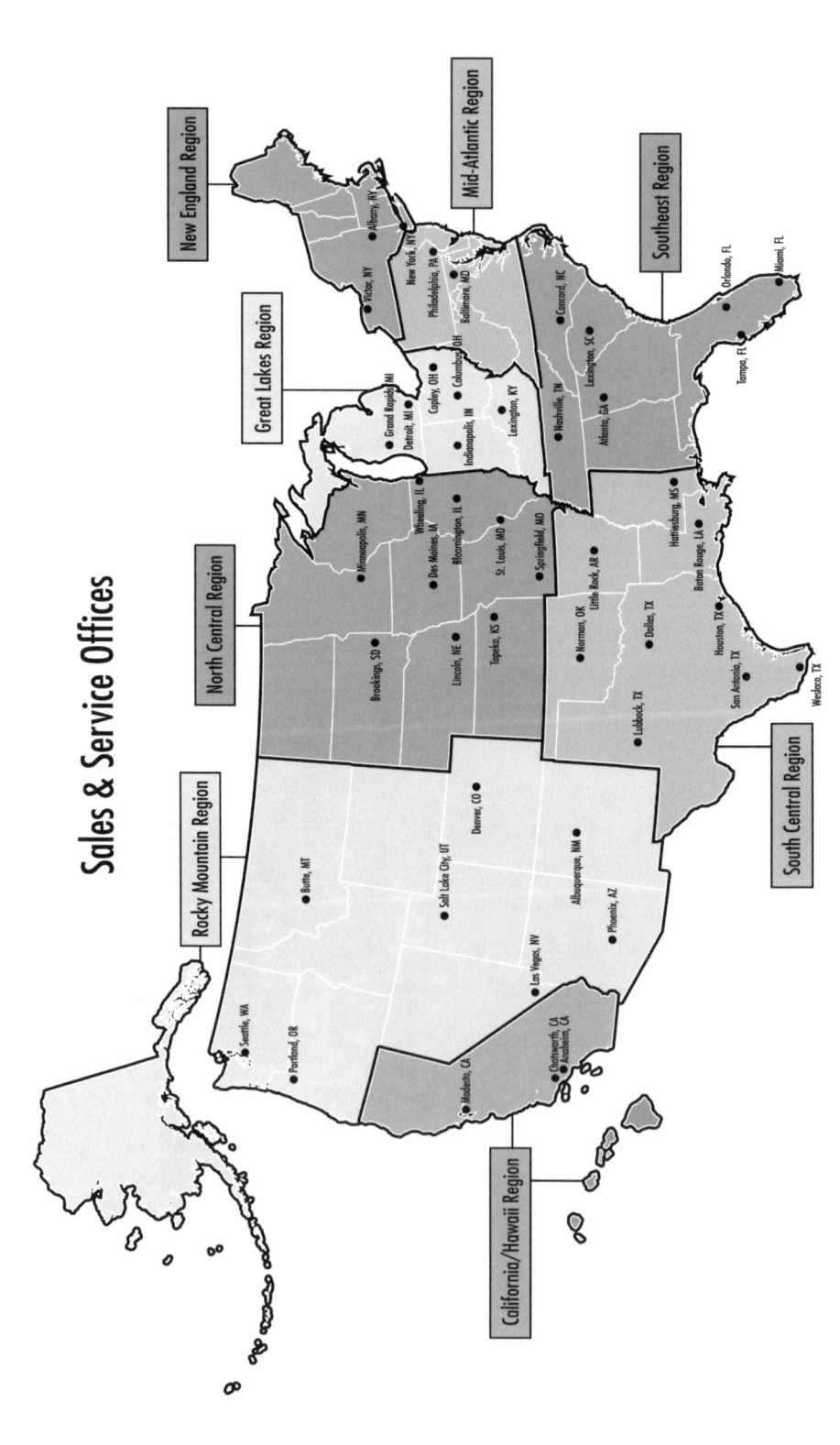

Rocky Mountain Region

Arizona

A One-Man Show in Phoenix

Most managers who have opened a Daktronics Sales and Service office were chosen to venture into a new market all alone; Mike Mayhew is no different. Mayhew talks about opening the Phoenix, Arizona office in the late fall of 2000.

"I was a one-man show. I sold, did the paperwork, and cleaned the bathroom. It felt overwhelming at the time opening the office in Phoenix."

In an effort to gain a foothold in the Phoenix area Mayhew didn't settle down to one market. "I've had the opportunity to work in every market segment. We never had a true market specific sales person in Arizona until last year. I assisted in all markets because there was no one else out here to support them. My primary focus was HSPR for the most part, but I assisted in anything and everything."

Mike Mayhew in the Phoenix, Arizona office (2000).

Working through the difficulties of a new area and working in multiple markets, it took five years for the Phoenix office to reach a milestone in their history. "The first major goal was to reach one million dollars in the High School Parks and Recreation market which we were able to do for the first time in 2005."

The Importance of Daktronics' Local Presence

"I think that most customers are greatly impacted from our local presence; which is the main reason we have been and continue to be successful. If it weren't for that, we wouldn't have the market presence that we do today. I think we are starting to see that in our international market.

Phoenix sales staff

We've purchased companies and opened offices. As soon as we have people on the ground, printing our product, we see market growth," said Mike Mayhew.

Daktronics' local presence not only helps with sales and service, but is important for installation support for difficult projects. One such project was the Phoenix Sky Harbor Airport that Mike Mayhew helped install.

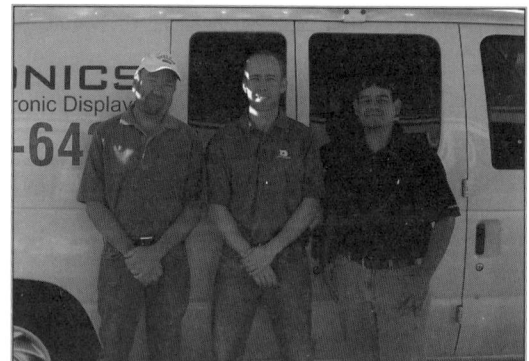

Phoenix service staff

"We currently have around 50 displays running at US Airways gates. To install the displays we had to do the majority of the work at night, due to airport regulations. I wasn't that familiar with the Galaxy® product and it's not that easy to call the help desk in the middle of the night. It was quite challenging to say the least."

The importance of Daktronics' local presence also shines through when building relationships with current and potential customers. "I have ties to my first sell ever at South Mountain Community College, one of the campuses in the largest community college districts in the United States...

Sky Harbor International Airport

It was a set of shot clocks. The reason we received the order was because I walked in and talked to the Athletic Director. We have a relationship with the entire district today and they rely on us for all scoring needs. It's a stand out because it ties back to the very first sell. The Athletic Director is still there and she has nothing but good things to say about us."

Colorado

Managing a Large Cliental Base in Sheridan

Scott Dieck opened the Sheridan, Colorado Daktronics Sales and Service office during the spring of 2002. The Sheridan Daktronics Sales and Service office is nestled in the Rocky Mountains and is south of Denver. The office expanded from Scott to a staff of twelve servicing a large cliental base.

"We opened the office and went to $325,000 sales and we're anticipating over a million this year. The staffing went from just myself to a staff of eight, now we have a total staff of 12 that cover the Colorado territory. We maintain equipment for HSPR up to the LSVs in town which includes; Coors Baseball Field, Invesco Field, Mile High Stadium, and Dicks Sporting Goods Stadium (professional soccer field). In the last year, we've been able to sell three Sportsound® systems, our first Galaxy® Pro for a high

Brighton High School – One of the many Lehi, Utah, projects.

Scott Dieck at his office in Sheridan, Colorado.

Colorado Rockies scoreboard during the 2007 World Series.

school, as well as selling the Junior College World Series. Grand Junction Park and Recreation also purchased equipment to be put on their field. We've serviced HSPR customers and professional customers with a large customer base we service and maintain."

Idaho

FEB 2 8 2008

Grace High School

704 South Main • P.O. Box 348 • Grace, Idaho 83241
Stephen C. Brady, Principal   Richard Condie, Athletic Director
sbrady@sd148.org   rcondie@sd148.org

Phone: (208) 425-3731
Fax: (208) 425-3063
www.sd148.org

Fighting Grizzlies

Feb. 25, 2008

Mr. Jim Morgan, CEO
Daktronics, Inc.
Brookings, SD

Dear Mr. Morgan,

I want to take this opportunity to tell you how pleased we are with our new Daktronics scoreboards. We were able to purchase them through a grant from the United Dairymen of Idaho and along with the hard work of your Idaho representative Bob Holsclaw, we now have new boards to replace the 30 year old boards we used to have.

I wanted to tell you how much we appreciated Mr. Holsclaw and his willingness to work with us. As I recall it was no more than thirty days from the date of the order confirmation to our scoreboards being delivered. Bob was in touch with me every step of the way and answered any questions I had. Along with that, Bob followed up and made sure that we had everything we needed, for which we are grateful. Bob couldn't have done more for us, and, in fact, went "above and beyond" to make sure we were satisfied.

Thanks again for what Bob and your company have done for our programs here at Grace High School.

Sincerely,

Richard Condie

Richard Condie, Athletic Director
Grace High School

Home of The Warrior

NOV 2 7 2006

West Minico Middle School
155 S. 600 W.
Paul, ID 83347
Ph. (208) 438-5018
Fax (208) 438-8513

Minidoka County Joint School District #331
· Academic Excellence
· Ethical Behavior
· Personal Responsibility
· In a Caring Environment

November 22, 2006

Jim Morgan and Gary Graham

Gentlemen,

We would like to take a few minutes of your time and let you know how refreshing it was to work with Cory Maxson and Robert Holsclaw in our recent purchase and installation of our score board. Through the entire process, Mr. Maxson and Mr. Holsclaw acted in a professional manner. When we discovered that the "lift" that we were going to use in the gym didn't fit through our auditorium and we had to use scaffolding, Cory and Bob didn't skip a beat and installed the score board without any trouble. You are indeed fortunate to have employees such as these working for you.

Sincerely,

Sandra Miller, Principal

Michael W. Hale, Assistant Principal/Athletic Director

OCT 1 5 2007

LIGHTHOUSE CHRISTIAN HIGH SCHOOL
259 MAIN AVENUE EAST, TWIN FALLS, ID 83301
208.737.1425   208.737.4671 FAX
WWW.LIGHTHOUSECS.ORG

October 12, 2007

Jim Morgan
Daktronics Inc.
Brookings, South Dakota

Subject: <u>Thank you</u>

Mr. Morgan,

My name is Nick Karavedas. I serve as Athletic Director and Head Football coach here at Lighthouse Christian School in Twin Falls, Idaho.

We would like to pass along our appreciation to you and Daktronics for the excellent service provided to us as we launched into establishing the "Lighthouse Athletic Complex." Bob Holsclaw worked with us in identifying our needs and walking us through the entire process. He was our liaison as the United Dairymen of Idaho donated two basketball boards and worked with me every step of the way to design our football board to compliment our new, start of the art, artificial turf football field. His knowledge of the product and experience in athletics were invaluable.

Again, thank you for providing an excellent experience. We look forward to working with your organization in the future.

All the best,

Nick Karavedas
Athletic Director
LCHS

"HOME OF THE LIONS"

Montana

Face-to-Face in Butte Gains Sales

Danny Defries opened his home office in Butte, Montana, in 2004. After working with Daktronics for 11 years, Danny admits it has been a challenging experience, but he finished his last few weeks in March 2008 without regret.

One of Danny's most significant sales was at Montana Tech in Butte, where they installed an entire full-color system which was funded by Pepsi® (a major sponsor for scoreboards throughout the country). After experiencing the great service and reliability of the boards, Pepsi® was sold on Daktronics' products. The great performance gave Pepsi® the confidence to decide to buy nothing but Daktronics' boards. Pepsi® admitted that Defries was the only salesman for scoreboards that ever came in their door to sell them a product.

Danny Defries

Through this partnership, Daktronics boards are in every community, tech, and state college in Montana. That is quite an advance from only having boards in two of the schools in Montana. Word of mouth, so to say, created a lot of sales.

Daktronics was able to expand in the area because schools continually wanted to upgrade to a Daktronics display. The schools saw new boards going up in other schools and realized the improvement in the way the product worked.

Montana Tech – Home of the Diggers

"Everyone sees it and wants it," said Defries about the spreading trend of Daktronics boards in area schools.

Building relationships was very important for Defries. He built most of his sales from face-to-face relations with the customers. He would pass through small towns and stop to talk to schools on a regular basis.

Defries was also a great service technician. He would receive phone calls about small problems with the Daktronics' boards. These problems ranged from the board not being plugged in to the time on the operator's computer being wrong.

Basketball Scoreboard at Montana Tech of the University of Montana (2007).

One school called Defries on a Friday afternoon before a game because they couldn't get the board to turn on. Defries listened to the problem and was able to solve it over the phone. It turns out, the board's automatic shut off timer was set for 5 p.m. every day and their machine was simply powering off.

Most of the problems that Defries dealt with were merely small quirks that could easily be fixed. Defries jokingly said the only real problem he had with the Daktronics boards was that, "The boards last forever so you can't sell to that person again."

Since it opened in 2004, the Butte office has been especially successful. After securing the contract with the Montana Department of Transportation (DOT) in October of 2007, Defries was able to exceed the sales goal by $100,000, reaching a total of $300,000 in sales. This project was special to Defries because he had been trying to secure it for nine years. When he first moved to the region and opened the office, he was able to get acquainted with the department. As soon as the competitor's board malfunctioned, he stepped in and closed the deal.

When driving by the DOT board one day Defries noticed a few of the modules were not working correctly. Since he had a key to the display, he decided to stop and see if he could fix the problem. He spotted a lot of the wires were hooked-up wrong. Baffled why the wires were messed up when he had only installed the board a few weeks ago, Defries believed it was his duty to try and fix the problems. Afterwards, he contacted the DOT only to find they had instituted the problems purposely in order to test new software that would find problems such as disconnected wires. The DOT was baffled at how the board had "fixed itself." The problems they created were not supposed to be noticeable to traffic passing by, but since Defries is an expert on the boards, he had noticed them. This only proved to the DOT the extent of his knowledge of the product. Despite the hindrance on testing their software, the department was glad to know someone was watching over and was amazed at the follow-up service offered without request.

Scoreboard News

The Value of Local Service
by Nikole Muzzy

Danny Defries

Daktronics has made it a priority to provide our customers with local service. There are Daktronics offices in more than 45 cities across the United States and a network of independent dealers who are there to support customers long after the sale. Danny Defries with the Daktronics Sales & Service office in Butte, Montana is learning just how valuable that presence is.

The Butte Legion Baseball team needed a new scoreboard and Harrington Pepsi agreed to purchase the board, in exchange for advertising space. Defries set up a meeting with Jim Bennett of Harrington Pepsi and the manager of the baseball team to go over different scoreboard options. The manager of the team did his own research on scoreboards before the meeting and decided he wanted a competitor's board because of the low price.

Before Defries could respond, Bennett told the team's manager Harrington Pepsi would only buy a scoreboard from Daktronics. "I have a good relationship with Jim and Harrington Pepsi. When he puts his company's logo on a product, he wants to know it will work. He knows Daktronics has quality scoreboards and if something goes wrong, he knows I will be there to help. Jim really appreciates the DSS office in Butte," said Defries.

Defries started building the relationship with Harrington Pepsi when he first started working in Butte. It did not take long for Bennett to realize that Daktronics could provide something other scoreboard companies did not offer in Montana – local service. Defries said, "Jim won't haggle with me over prices because he values my service. The Butte Legion Baseball team is just one of many orders I have received through Pepsi. And he and I are working on several more."

Another memorable sales transaction for Defries was actually an accident. After a great sales transaction and on time delivery, Defries was on-site to witness the delivery crew run a forklift into the side of the scoreboard. The board was completely destroyed in front of the high school officials. Luckily, Defries was on site to get a new order placed. Within a month the school received the new board in time for their first game of the season. Because Defries was able to manage the problem smoothly, the school felt reassured and didn't worry about it further.

Nevada

The Bright Lights of Vegas

In September 1998, Jo Gasper and Steve Weldner moved from the corporate offices in Brookings, South Dakota to open the Daktronics Sales and Service office in Las Vegas. When people think of Las Vegas they usually think of the bright lights on the Las Vegas strip, where 18 out of the 25 largest hotels in the world are located. Daktronics' presence on the four mile stretch of Vegas is undeniable. Currently, Daktronics' has eye catching displays on the strip including: Venetian® Hotel and Casino, Gold Coast Hotel and Casino, New York–New York Hotel and Casino™, the Palazzo, and Mandalay Bay.

Gold Coast Hotel Casino

New York–New York Hotel and Casino™

The Venetian® Hotel and Casino

The Venetian® Hotel and Casino

Janet Van Ess, office Coordinator, says that not all projects are as glamorous as the strip's, but they still play a pivotal part in Daktronics' foothold. "One that I'm working on right now is with a park that's looking to order 12 baseball scoreboards. The project for the new park has been in the works for almost six months...working with this customer led me to two other sales and I'm waiting for the third sale to come through."

Sony® PSP display at the showcase Mall.

Janet Van Ess

For three years Eric Johnson has worked at the Las Vegas Daktronics Sales and Service office as a Sales and Service Engineer taking care of Daktronics' many customers on, and off, the strip. "We're more proactive than reactive. We watch the customer sites and solve issues before they happen. This helps reduce the amount of overtime we work. A continuing goal is to maintain relationships between arenas and casinos in Vegas."

Taking Service to New Heights

The Palazzo's marquee is over one hundred feet high. Janet Van Ess talks about the difficulties that are included in working with a display at such a high level. "Technicians have to climb it to get to the top. So, if they forget their tools they have to climb down. They can't have a boom lift because of where it is."

Eric Johnson knows that fixing the sign is a high priority since it is at such a prominent location. "It's a long climb, but it's really not that bad, it's the fact that it's up a tiny tube. There are a lot of power supply issues. About two weeks after we installed the Palazzo marquee we started having power supply and power issues," said Eric.

Typically, the team is out there within a day. The Palazzo has a 16 hour (two business day guarantee). Daktronics was there within the allotted time period. But, it took a day longer to get the right part in the display. Even though it took another day, Eric Johnson says the relationship between Daktronics and the Palazzo is solid.

The Palazzo display, Las Vegas' newest Resort-Hotel-Casino, located on the strip (January 2008).

"They constantly get remarks about the displays. 'Those displays are the best displays on the strip.' We have a very good relationship with them."

Blue Man Group

Another Daktronics installation in the Venetian® is the Blue Man Group's stage. Eric Johnson says that without their equipment running properly the show is not going to run.

"The Blue Man Group was a fairly detailed installation. We set up six ProStar® dis-

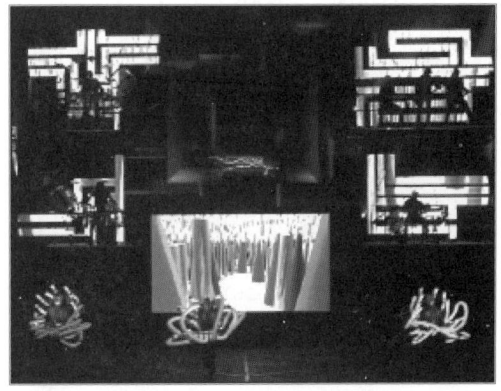

The Blue Man Group's main display in the Venetian®.

plays that looked like a crossword puzzle. Set up their live video and animations...the whole nine yards. That was my very first install...it was a very big learning experience trying to integrate all that equipment. The displays are a very large part of their show. If the displays go down...they're done. They have never had a problem since it's been installed."

Daktronics Engages Competition in Shoot Out

For 24 hours, Daktronics HD-16 technology stood against the competitor's 20 mm display on the 80 acre construction site of the future M Resort. Marnell Corrao Associates, a large architecture firm in Las Vegas responsible for the construction of Treasure Island, Wynn Las Vegas, Bellagio, Mirage, and Rio Palazzo Suites (to name a few) placed the signs next to each other in an old fashioned shoot out, reminiscent of the wild west that Las Vegas is known for.

The companies set up the shoot out to help Marnell Corrao Associates determine which displays they will use for the billion dollar resort, scheduled to open Spring 2009. To keep the competition fair, Keyframe® provided food and restaurant content for both displays to run. This way, the customer could see which display out-performed the other in day and night-time conditions.

"The customer comes by and takes a look at the two running next to each other and they make their pick based off of image quality," said Matt Evers.

Matt Evers

NEW MEXICO

"Service after the sale" Keeps Albuquerque Office Running Strong

In 1999, Randy Ramsbacher started the Albuquerque, New Mexico, office. Since then, the office has had nine very successful years. Fiscal year 2008 was especially profitable, with the office exceeding its sales goal of $300,000.

Dan Gallegos is currently the head of the Albuquerque office. For the past seven years, Dan worked intermittently beside various office coordinators in Albuquerque. Not always having the benefit of a full-time office coordinator on staff, he often worked on projects by himself. Recently, the office has experienced growth with the addition of a new technician to its staff.

Dan has made a tremendous impact in the Albuquerque office since he started. He has provided Daktronics displays to almost all the venues in the Albuquerque area with completely new products and has even replaced many of the competitors' old boards.

Specifically proud to be able to work with the University of New Mexico, Dan Gallegos assisted the university in upgrading its displays. The University of New Mexico started with some small Daktronics products. Now, the Daktronics logo can be found on every video board and scoreboard at the university. Dan has kept in contact with the university on a service level basis, resulting in improved confidence in Daktronics' products and service.

Dan Gallegos

When the University of New Mexico was asked what

University of New Mexico.

made them switch to Daktronics boards, they credited their decision to the top notch service that Daktronics is known for. The ability to have local service people on-site has been a huge factor in the continued success of the partnership.

The Albuquerque Public School is also another noteworthy project for the Albuquerque office. For years, Daktronics was only able to service other manufacturers' boards within the school system. Eventually, Dan Gallegos was able to step in and make the sale for Daktronics. Now after three years of dedicated work, Daktronics has six displays on its football field alone. The school sports a large football scoreboard, a RGB marquee, and a Sportsound® 1000 on each end of the field. Daktronics has supplied every new scoreboard the school has requested since.

Building from the fact that Daktronics has an excellent product to sell, the company has continually supported its product with continued service after the sale. While having confidence in the product is key, backing up the product is huge. The dynamics of the product are what have allowed Daktronics overtake this region.

Albuquerque Public Schools.

A hot air balloon passes by an Albuquerque, New Mexico, Daktronics Sales and Service van.

Oregon

Portland Office Dedicated to Customers

Working with a strong team since 2000, the Portland office has been successful in working on many niche markets.

By focusing their effort on aquatics, the team was able to successfully work on a market with no competition. Daktronics is currently very strong in the aquatics market with the implementation of the new Relay Take-Off Platform that is rivaled by no other company.

Daktronics has a good history of track timing installations, especially when it comes to interfacing with scoreboards. Cottage Grove High School was a highlight project for the Portland team because they were able to install Daktronics boards for every scoreboard on their campus including a finish timing system and a Galaxy® at the front entrance.

Leading the Portland team, Steven Davis was able to make a significant impact in the Portland area. Going back to his roots, Steven Davis sold Linfield College, the college he graduated from, its first ProStar® and Sportsound® outdoor football board.

"On this project, I was particularly proud to sell to a school where I am an alumnus," said Steven Davis, "It's nice to go back to a place and faces you recognize."

Cottage Grove High School

After installing a 20mm RGB display, a car dealer noted within the first nine months of installing his board his sales had increased. He gives the new installation credit and says it was the best advertising money he had spent.

Through positive feedback from customers and extensive knowledge of the product, they were able to create good relationships with their customers and business partners. When a school called saying their board was showing the home score on the visitor side, Davis went to the school to fix the problem himself without telling them how easy it was. Once explaining the problem they all laughed. This proved that he did in fact know the product well. This made the school very comfortable with Daktronics' service and has since brought in more sales from them. Their dedication to their customers shows through all of their repeat business.

Steven Davis

"Every service call is a sales opportunity. If we take care of the customer on the service side then we will be successful on the sales side. We work hard to find customers and we don't want to lose any of them," said Steven Davis.

Their dedication and support extends to co-workers too. When Chris Wagoner from the Modesto office called in need of a part soon, they planned to each drive half of the seven hour drive to get the part. Due to snow they were not able to accomplish this, so Davis flew out at 9:30 that night and headed to San Francisco with the part in his carry-on. When he got there he handed off the part. After spending the night on an airport couch, he flew back at 7 a.m. to return home. He realizes how important not only his customers are, but the customers throughout the business to repeatedly prove their dedication to all customers.

They pride themselves in being able to cross-market support for the various business groups and the great relationship they have developed. Through these relationships they have been able to gain a solid base with all college levels, Division one through three.

Utah

Success Stories in Lehi

The office in Lehi, Utah, (located south of Salt Lake City) was opened during the summer of 2001 and hasn't moved since. It went from a one-man show, to a four man production. Cory Maxson knows that relation-

LEADERSHIP THROUGH SERVICE, A PHILOSOPHY THAT SELLS

By Barbara Kleinjan

With just a quick web search of Linfield College in McMinnville, Ore., it's easy to find Steve Davis' name. A prolific kicker for the Linfield Wildcats, Steve was inducted into the college's Hall of Fame in 2003. His former coach and longtime friend, Ad Rutschman, called Steve, "a tough competitor who was great under pressure."

Steve Davis isn't kicking footballs any more, but his competitiveness remains. Frank Kurtenbach, Vice President of Sales for Daktronics said, "When I think of Steve, I think of his love for sports. I like his enthusiasm for life." The Portland Daktronics Sales and Service office follows their leader's passion to take care of their customers.

"I've been one of Steve's customers and now I work for him," said Lynn Freshour, a former athletic director turned DSS salesperson. "He's like a player's coach. He is a true leader who gives of himself first and that makes me want to work harder for him."

True leadership serves. Serves people. Serves their best interests. Steve Davis and his staff have established a model of customer service based on the principles of leadership, lessons learned when Davis played college football. With a sales district encompassing three counties around Portland (nine hours wide) and including 400 high schools, 14 colleges and four NCAA Division I universities, Davis and his team of nine employees take their mission seriously. Focused on the customer, Davis trains his co-workers to build relationships. Lynn said, "My job is easy because I work for Steve. All across the state of Oregon, when you mention scoreboards, people think of Steve Davis." He takes care of his customers and people like that."

Graduating from college with a degree in journalism and sports media relations, Steve's first career strengthened those skills. He joined Daktronics at the age of 50, knowing that he "took a leap of faith," supported by his gray hair, past experiences and ability to see the "big picture." His attraction to Daktronics was not a financial one, but rather an appreciation for the employees of Daktronics and what they stood for: integrity and a human approach to business.

Take care of people and they will reward you. Coaching his Portland office as he would a championship team, Davis' district has seen tremendous growth. As a creative and circular thinker, he calls his plays instinctively, hoping he has trained his employees well enough to switch direction when the customer's needs dictate it. As the "face of Daktronics" in the Portland area, Davis realizes that the team is a mirror image of the coach. "I like people. I enjoy making people happy when we work for them. We work too hard to acquire customers; we don't want to lose them."

Steve understands that if the coach does well, the team achieves success. He knows that "ultimately every loss is the coach's fault: lack of fundamentals, lack of preparation, lack of motivation, lack of ownership. We win as a team and lose as a team. If we break down or make mistakes, we must find a way to come back stronger and better. The next time we must do it right. We communicate extensively with our customers; we ask our employees to brainstorm better solutions; we examine all situations as if breaking down game film." Mark Gomez, President of the Le Grande Swim Club saw Steve at work and within one hour, was sold on Daktronics service. (See Gomez's testimonial on page 9.)

Every service call is a sales opportunity. Solving the customer's problems has become the motto of the Portland office. Steve and his team will answer the phones on Sundays, take service calls for emergencies, drive through the night to replace a timing part, or do whatever it takes to help their customers achieve success. Steve enjoys selling Daktronics equipment because "the quality of the product surpasses the competition. It is fun to be number one." He's a tough competitor and he's still at the top of his game.

Steve Davis is the Office Manager of the Portland Daktronics Sales and Service office

Lynn Freshour
HSPR Sales

Stan Gila
HSPR Sales

Amanda Larson
Office Coordinator

> We communicate extensively with our customers; we ask our employees to brainstorm better solutions; we examine all situations as if breaking down game film.
> - Steve Davis

ships, between Daktronics and its customers, are vital to gaining customers and result in many "success stories."

"Up north we deal a lot with Pepsi®...Bonneville High School has become a success story for Daktronics. Brad Houston built up the relation-

Cory Maxson

Brighton High School – One of the many Lehi, Utah, projects.

ship with the Pepsi® bottler. The Athletic Director needed some new scoreboards to replace some old, rundown scoreboards. We've sold them two baseball scoreboards, a basketball board in the gym last year, and this year we'll install a new football scoreboard; replacing all competitor's equipment."

Awkward Installation

Every person who has ever installed a scoreboard knows that sometimes, the conditions aren't ideal. Maxson knows this first hand.

"In Paul, Idaho, we installed a basic basketball scoreboard in the West Minico Middle School in Paul. One of Daktronics' Retired Athletic Directors, Bob Holsclaw, sold the board and we were going to do the install. We couldn't use the lift so we had to install with scaffolding, which was just your standard board on a small little walkway. I ended up being the main person doing the install while Bob Holsclaw held the scoreboard. It was quite interesting."

Unexpected Visitor

Marlo Jones visited Lehi, Utah, on February 12-14, 2008. Without this trip, Cory Maxson and Marlo would have missed out on quality bonding time.

"The night of the 13th there was a major snowstorm that came through the Lehi, Utah, office area to the point that we got snowed in and couldn't leave the office. The interstate was at a stand-still. There were only a few inches of snow, but wind gusts were up to 60 miles an hour, thus making huge snow drifts. The hotel was about 15 minutes away, but all the roads were closed. We ended up staying in the office until 11:00. Marlo cleared off a desk, was going to lay on it, and use bubble wrap as a blanket. Then it got to the point where we could travel on the road. I talked Marlo into spending the night at my sister's house, a mile and a half away."

WASHINGTON

The Very First Daktronics Sales and Service Office in Seattle

The Seattle, Washington, office was opened by Marlo Jones in 1988 and was the very first Daktronics Sales and Service office (then called Scoreboard Sales and Service). Twelve years later, Kyle Williams was hired as a salesperson and technician. The Seattle office works with high profile customers, such as: Key Arena (Seattle Supersonics [NBA], Seattle Storm [WNBA], Seattle Thunderbirds [WHL]), and SAFECO Field (Seattle Mariners [MLB]). The first time the office hit the million dollar mark was during the fiscal year 2001, and they haven't been under the million dollar mark since. As a matter of fact, the office hit the two million dollar mark in 2005 and 2006. Kyle Williams says part of the success is from the willingness to stay late into the night and finish projects for their customers.

"Key Arena was really my first exposure to our video products. We installed, back in 2001, a 360 degree ProAd and got it up the day before their first preseason game. If I were to go

Key Arena, Seattle Washington

back at the paperwork and look it up, I remember staying there until eleven o'clock every night for a week, usually, after spending a day in the office."

Daktronics' employees also exhibit the humility to clean filth ridden signs. Kyle Williams talks about how he got the "opportunity" to clean a layer of dirt off of a scoreboard that traveled across half the country from Brookings, South Dakota, to Seattle, Washington.

Kyle Williams

"SAFECO Field was finished mid-season in 1999 (before I started with the company). But in 2002, we replaced the video board there. We replaced it with the exact same thing that was just removed. In January that year, we replaced the video board and the funny part was that when it showed up it hadn't been covered correctly, so it had several thousand miles of road dust. I worked with another employee to scrub the entire face of the sign. The installation itself went off without a hitch in January."

SAFECO Field

Piece by Piece

Installations are part of everyday business at Daktronics Sales and Service offices. Some installations are more difficult than others, especially when the install crew can't use any lifting equipment. Kyle Williams experienced such an install at the Mercer Island Beach Club.

"A couple summers ago I worked on a job at the Mercer Island Beach Club (private pool on Mercer Island, in the middle of lake Washington). They were doing new construction, basically, building a brand new facility. They have an outdoor six-lane pool and we did all of their equipment for them. We were working directly with the club during a difficult coordination project. Luckily, we had a lot of students working with us in the summer of 2006. That helped us put the project together without many hitches. When all was said and done our student labor stepped up to the plate and helped us install the project. Everything was done on ladders and with our hands. They didn't allow any lifting equipment on their pool deck for fear of cracking the brand new concrete. It was all done by hands and ladders. The overall height of the scoreboard was 5 feet 10 inches high, and 27 feet wide. It all came in pieces. Piece by piece we carried them up the ladder and bolted them to the mounting structure. I think we were on-site for about three days. The customer was very satisfied. I do remember him telling us that out of all the contractors he worked with on the project, we were the best one he worked with. It was a lot of work and wound up during our one nice week. It was good that it wasn't raining but bad that it actually got hot that week."

After working on such a tough installation, Kyle Williams was asked to turn around and work on another project. Without a second thought, he went to Little Creek Casino.

"The Mercer Island project happened to coincide from a commercial project. I got a call from a company (about three hours away) whose sign wasn't up and running. Little Creek

Little Creek Resort and Casino

Casino had a grand opening that Friday night. I got a call from them at three in the afternoon and of course we were in the middle of installing equipment. After the install, I went right down to the casino. It took about three and a half hours to get there. When I got there it was dark. It was a matter of getting the sign up and running by the light of the sign. We got it working that night; I left the sight at 10:30 and it was working. If I remember correctly Bill Cosby was their performer that weekend. It was on so he could see it when he got into town that Saturday."

Freeze Up

Before moving to Kansas, Jim Preston worked at the Seattle office from 1995 until 1998. As a new service technician with Daktronics Jim was asked to keep the Seattle Supersonics' display running for a nationally televised game.

"I worked in Seattle for three years. When I was brand new I worked to get the Seattle Supersonics' display running. We were contracted to be there for the first bunch of games. They put me out there for a televised game in case something went wrong. They tip off, the clock starts, but less than three seconds into the whole game, the clock freezes up. I had to walk out to the center court and reboot the system. I got some 'Booo's' and 'What are you doin' out there?' But, it worked fine for the rest of the whole game."

Seattle service van (1996).

Jim Preston working in the Seattle office.

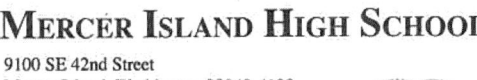 MERCER ISLAND HIGH SCHOOL

9100 SE 42nd Street
Mercer Island, Washington 98040-4199
(206) 236-3345 • FAX (206) 236-3358

John Harrison, Principal
Craig Olson, Associate Principal
Michael Schiehser, Associate Principal

February 9, 2006

I wanted to let your company know about the great customer service I received from the Seattle Branch of Daktronics this winter. I am the host for a high school league championship swim meet – a competition which includes over 200 athletes. During the preliminaries on February 3rd, my Omnisport 6000 began to act strangely. After much effort – I had to stop using the system that evening.

After the preliminaries, I made a phone call to Joe Richer, who directed me to Kyle Williams. Kyle was helpful while on the phone, directing me through a number of tests in order to come up with the best answer he could for what the problem could be while I had him on the phone. Kyle then went to the Daktronics office at 10 pm on a Friday evening in order to find another 6000 that worked so that I could use it on Saturday for the finals of my meet.

Not often in this day and age does a consumer find a company willing to go the extra mile for them after purchasing a product. Daktronics is not like that. I appreciate their willingness to work with their customers to keep them happy, and keep their equipment working.

Thank you for your efforts on behalf of all your customers in the Northwest.

Sincerely,

Jeff Lowell
Head Swim Coach
Mercer Island High School

California/Hawaii Region

Anaheim

Pulling Together as a Team in Anaheim

On July 3, 2003, Rocky Gruner started the Daktronics Sales and Service office in Anaheim, California, in a 2,000 square foot office complex. Today, the office is thriving and has "upgraded" to a 4,400 square foot office space. There are currently 14 employees based out of the Anaheim office.

Gruner recalls an interesting situation when he started the Anaheim office. In 2003, National Sign and Marketing Corporation was an avid competitor for Daktronics in the California market. Through Gruner's persistence and hard work, National Sign and Marketing Corporation now is a reseller for

Rocky Gruner

Angels Stadium (Spring 2004).

Daktronics. Gruner said, "I used to be a competitor with people who are now my customers."

In 2005, Daktronics sold a display system for the Angels of Anaheim stadium in Los Angeles. One Friday morning, Gruner received a phone call from Jim Morgan and Fred Doremus, the Project Manager for the contract. Morgan and Doremus needed Gruner to round up around 25 employees to fix a technical issue with the sign. Gruner and 25 sign installers met at the stadium on Friday and worked through the day and into the morning until 2 a.m. The next morning at 6 a.m., the installers returned to the stadium and worked until the game at 7:30 to fix the issue. With effort, teamwork and late hours, the group was able to get the display up-and-running in time for the game. Gruner explains that they worked hard, "so the fans at the Angels stadium would have the experience that Daktronics promised. We pulled together as a team to pull it off."

Anaheim and Chatsworth Staff. Back row (left to right): Karen Kracji, Mike Gibbons, Kathy Parsons, Al Davino, Rocky Gruner, Rusty Van Cleff, Ron Vaught, Bob Frechner, and Carl Buendia. Front row: Israel Robles, Jesse Donini, and Juan Carlos.

Anaheim Office: Back row (left to right): Rocky Gruner, Kevin Sosa, Jee Yoo, Ted Alcorn, Isreal Robles, Charles Rendon, and Rudy Papaco. Front row: Louisa Trevino, Dan Bati, and Lori Hensley.

Chatsworth Office: Back row (left to right): Marcos Espinoza, Michael Gibbons, and Ed Wasserman. Front row: Jee Yoo, Karen Krajci and Carl Buendia.

GIBBONS GIVES HIS BEST

By Danylle Rozier

Mike Gibbons (above) is a sales and service representative for the High School Park and Rec located in Chatsworth, California.

Daktronics Sales and Service representative Michael Gibbons doesn't waver when it comes to the customer service he provides. He believes in providing outstanding service that benefits the customer and fulfills their needs.

Gibbons, who manages Daktronics Chatsworth, Calif., office, has been a member of Daktronics Sales and Service team since October 2003 and has 25 years of experience working in the aquatics world.

Gibbons, a graduate of Liverpool College of Technology in Liverpool, England, believes in being hands-on with his customers before, during and after their events. "I like to attend events to make sure our customers know how to use the equipment during competition. It is different using the equipment in a non-competitive setting than it is to use the console during a competition," said Gibbons. "It doesn't matter if it is several weeks later, a number of meets later or even if we have already trained the customer on the use of the console."

Gibbons' dedication is reflected in the comments Daktronics hears from his customers. "I cannot say enough about Mike's help in getting this deal done (See Fullerton Union High School, article) and getting the system up and running. From the beginning he was supportive and collaborative. His support during the pre-installation preparation, installation, training and follow-up were absolutely unparalleled," said Dan Notti of the Fullerton Union High Aquatics Booster Club.

Pat Hagele, Recreation Services Administrator for the Santa Clarita Sports Complex Aquatic center, agrees. "We have had excellent support in all ways from Daktronics — especially from the company's local office and from Mike Gibbons in particular," said Hagele. "With our Daktronics timing system, along with Mike's dedication and commitment to providing exceptional service, the aquatics center has been able to effectively host the type of events it was designed for, achieving a long-term city goal of providing a high level of service to the Santa Clarita community."

Prior to working for Daktronics, Gibbons was an overall site project manager for Swiss Timing. He worked at the Atlanta and Sydney Olympic Games, where he was responsible for the preparation and infrastructure to support the electronic timing and scoring systems using Daktronics scoreboards for 36 sports in 22 venues. After the Sydney games, he served as a technical consultant for Daktronics for two years, covering Australia and New Zealand.

As the manager of the Daktronics Sales and Service Chatsworth office, Gibbons has particular responsibilities for the sales and service of Daktronics aquatics scoring and timing systems.

Daktronics Sales and Service Chatsworth, California

CHATSWORTH

California Office Continues Success

The Chatsworth, California, office was opened in December of 2000 by Don Hank. Three years later, Mike Gibbons took over the office. The Chatsworth office moved to a double unit in its office complex during 2007. There are six employees who work out of the Chatsworth, California, office.

Chaparral High School Aquatics board.

Gibbons recalls the selling of the aquatics system at Chaparral High School as a project that he was proud of. "We were able to give them what they wanted on a tight budget," said Gibbons. "The customer was highly delighted."

Modesto

Building Relationships Helps the Long Run in Modesto

Michael Gibbons

In 2001, the Hayward, California, office was opened by Chris Wagoner—the sole office employee. In 2003, the office moved to the neighboring city of Modesto, California, due to the increasing cost of living in Hayward. Currently, the office employs seven people.

Wagoner notes that he is proud of all of the projects that the office has sold and serviced, including the large and small projects. He is also pleased to see high school interest in video boards increasing.

Chris Wagoner

When selling, Wagoner has a philosophy of developing a rapport with the customer, "I am more interested in building a relationship because in the long run that will be more profitable," said Wagoner, "Basically what I try to do

Ryan Forunda, Chris Wagoner and Ed Koch (November 1, 2002).

is to sell what is needed—I go in as an expert and they tell me what they want and I tell them what I can offer."

While working on servicing or installing projects, Wagoner notes that he probably holds a unique record. He said, "While working at Daktronics, I have been hit by about every ball there is except a golf ball or bowling ball."

Another unique experience happened when he was working on selling a board to an old acquaintance—his technology teacher from the high school he attended. The teacher had lost Wagoner's business card and had to contact Wagoner's mother for his phone number. He admits that the teacher had his mother's number because of his unruly behavior during high school. The teacher frequently called his mother when Wagoner would skip class. In spite of this, Wagoner made the sale to his former teacher.

SUPPORT FROM THE PRESIDENT OF SALUKI SWIM CLUB, INC.

Clay Kolar, President
Saluki Swim Club, Inc.
620 Sheppard Lane Makanda, IL 62958
618-457-4627 bjkcak@hotmail.com

Dear Mr. Vargas:
As you are well aware, our swim club, the Saluki Swim Club, located in Carbondale, Illinois, recently purchased Daktronics timing equipment for our competition pool facility at Southern Illinois University. Items purchased included an Omnisport® 2000 timing console, 16 Lane Module IV's, extension cables and push-button back-up timers. We decided to convert our present electronic timing system to a Daktronics timing system for three reasons; one, the Daktronics display board we purchased three years ago has worked flawlessly; two, we did research on the various electronic timing systems available and found high customer satisfaction with the Daktronics equipment; and three, reports from references of customer service, follow-up and commitment to customer satisfaction were outstanding.

In addition to the above listed reasons, the Board and members of our Club and the coaching staff of the SIUC university swim teams appreciated very much your willingness to come to our facility and make a presentation of the timing system products you offer. The fact you brought samples of the entire system for us to examine and learn about was very instrumental in our decision to purchase from Daktronics. After our decision to purchase we were also very impressed and appreciative of your efforts to ensure quick delivery of the purchased equipment in time for our upcoming swim meet. Your office representative, Mary Meginnes, was very diligent in keeping in contact with me and following through with the order to make sure that all was correct and that we received all the items in time for our May meet. Furthermore, the availability and willingness of the technical representative from your St. Louis office, Bryan Nagel, to come to our facility prior to our meet and do a 'dry-run' to ensure a correct interface with our scoreboard and to instruct us in the operation of the console, was much appreciated and made a very positive impression on us. Indeed, your promise of customer service was delivered by you and Daktronics in all aspects of our purchase.

I will emphasize that your follow-through in ensuring delivery of the timing system so that we were able to use it for our well attended long-course spring meet in May was impressive and much appreciated. Due to the 'dry-run' experience prior to the meet we were able to conduct the meet without any timing system malfunctions, problems or delays. We quickly became very comfortable with the system and at this point we are very pleased that we chose to convert to Daktronics aquatic timing system equipment. Because of our satisfactions with the Daktronics timing system equipment we have purchased thus far and your delivery of customer service, as we build up our equipment fund we look forward to continue expanding our equipment inventory so that eventually we are still using all Daktronics touch pads.

We thank you for your time and effort in processing our order. We will be pleased to be listed as a reference for your aquatic timing system equipment.

Sincerely,
Clay Kolar, President
Saluki Swim Club, Inc.
Carbondale, IL

NORTH CENTRAL REGION

ILLINOIS

People Notice a Little Bit in Bloomington

The current Daktronics Sales and Service office in Bloomington, Illinois, opened in 2003. Jose Vargas knows that there's not just one project that makes Daktronics well known, but a consistent market share.

"The biggest accomplishment was super-ceding $100,000 in sales and this year we've hit the $200,000 mark. We're really proud of the University of Illinois; they have a bunch of video boards. With regards to high schools our proudest accomplishment is selling to Bloomington Central Catholic, it was the first High School scoreboard sold in Bloomington...There's not one thing that helps make us successful. A little bit here, a little bit there, people notice. Those are the projects that you are known for."

Jose Vargas

Perseverance and Patience

Being in the business for a couple of years Vargas has learned that not every service check goes as expected.

"Here's one that took me longer than normal. You couldn't hear the horn on the shot clock at the University of Illinois. For some reason, I

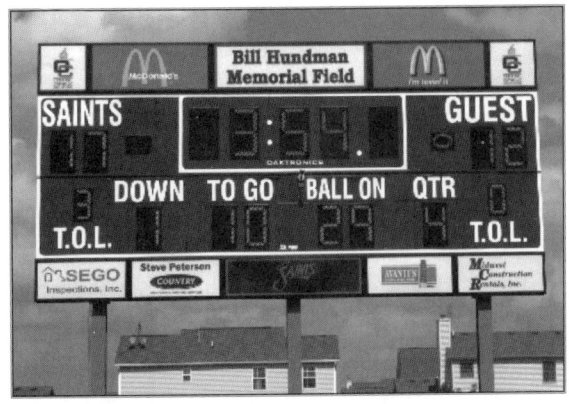

Central Catholic High School in Bloomington, Illinois.

thought it was one of the components. After fidgeting and swapping parts, I talked with a tech on the phone. He asked if I checked the dimming on the console. After I did that, the horn came back to life. It was a simple fix, but I had spent too much time on the repair."

Practical Jokers

Working at the Daktronics Sales and Service office in Bloomington, Vargas has learned to appreciate a joke, or two.

"The most recent was Lanphier High School in Springfield, Illinois. There was an issue with the scoreboard, so I dropped in to fix it. I visited the Athletic Director. He said, "you have some nerve walking in here (with a stern voice)." I was caught by surprise. 'We need to talk,' he said. I closed the door to the office. It turned out to just be a prank. He reiterated how grateful he was for the service he received earlier in the year. I'm sure I turned all sorts of colors as my jaw dropped wide open."

Wheeling, Illinois, Office Comes Through Under Pressure

The Wheeling, Illinois, Daktronics Sales and Service office opened during the fall of 2006. Located 30 miles northwest of Chicago the office does a lot of business with the world-class airports, Midway and O'Hare. Lorena De Avila talks about us how bad weather can cripple a vital display.

"A snow storm had been predicted to hit Chicago in the afternoon. Unfortunately, Chicago Midway was having problems communicating with one of their displays (that was operating off a Wireless Ethernet Bridge). The display was the only point where they could inform traffic of alternative routes and parking availability at the

Midway Airport during the Spring.

Chicago O'Hare international Airport indoor message center.

airport. Needless to say, they were in a rush to get this communicating before the storm hit. We received a call from the Airport requesting any possible help with the display to get it up and running. Luckily, our Field Technician, Scott Blankenship, was able to respond immediately and got everything running perfectly before the storm hit. I received a call from the Director of Landside Operation and Maintenance, thanking us for our immediate response."

Coming through in the clutch is extremely important while working with airports, especially since the Chicago airports are some of the busiest in the world. Lorena De Avila talks about how the priority increases in airports around the holiday season.

"It was Friday morning before Christmas and we were planning to install a display at Chicago O'Hare International Airport. Rich Krautter was providing installation supervision on the project; since it was just around the corner I stopped by to see how things were going. We were planning and hoping that things would move along quickly so that everybody could take off for the holiday weekend. Unfortunately, things were not going well (with the installation or the foggy-snowy-cold weather). Finally, we were able to provide power to the display. Upon flipping the breaker the power appeared to "pop." We then noticed that power had been killed to all the DMS, street lights, and Christmas decorations as you enter the interstate into O'Hare.

Roman Kreeger and Lorena De Avila.

This got the whole group pretty nervous, thinking that we just shut down power entering one of the busiest airports in the world, on one of their busiest days. After lots of scrambling around, power was restored, and all was well.

Wheeling Office. Left to right: Richard Krautter, Scott Blankenship, Lorena De Avila, and Kyle Sydow.

However, for about an hour we were thinking that if we didn't get power restored it wasn't going to be a very Merry Christmas."

Iowa

Helping Customers in Ankeny

From its start in 1990, the Daktronics Sales and Service office of Ankeny has surpassed its sales goals and is known for outstanding sales and service in the state of Iowa. There are currently three full-time employees within the office, and two retired athletic directors who are located in various cities throughout the state.

Of these employees, Darrell Thiner, the field Sales and Service manager in the Ankeny office, stands out among Iowan athletic directors and other common customers. In an interview with Barb Kleinjan from HSPR in the article, "Only as Good as the Service After the Sale," appearing in Daktronics' aquatic magazine, *The Latest*

Darrell Thiner in his Ankeny, Iowa, office.

University of Iowa

Score, Johnston Public School(s) athletic director Gary Ross states, "Do you know Darrell Thiner? You have to meet him! He's a great person!"

When asked about his personal philosophy on selling and servicing Thiner stated, "Selling, for me, has always been easy for me as far as talking to customers and making sure they know that we are there to help them. I think, 'What would I want?' We, as a company, need to put ourselves in the customer's shoes, and figure out what is the best equipment for them. We at Daktronics need to find out how the customers do things and provide for them exactly what they need and what's important to them."

The Ankeny office also proudly displays a reached goal of over $1,000,000 in sales during the year of 2007. "This is a number to be proud of considering we are eight miles away from Daktronics' largest competitor in Iowa; Fair-Play Scoreboards," says Thiner, "It's challenging to be so close, but we worked together and our sales both increased. To spread the word about Daktronics, I would talk to existing and retired athletic directors. We now have three major college contracts including the University of Iowa Hawkeyes, the University of Northern Iowa Panthers, and the Iowa State Cyclones. We gained a lot of sales and experience with these schools because all three schools' basketball equipment was installed in one year."

Thiner is also proud to have sold and serviced the first two video boards ever sold in Iowa at the high school level. Thiner played a part in these projects as the head salesman on one of the boards and an installer for the other.

According to Barb Kleinjan, Darrell gives credit to the many Daktronics employees stating, "Our equipment is superior and our service after the sale is unmatched."

Iowa State University

Kansas

Larger than Life Projects Give Daktronics Credibility in Lenexa

Jim Preston spent six years in the US Navy as an Electrical Plant Operator on the USS Hawkbill, Nuclear submarine. Then he attended South Dakota State University from 1989-1994. During his time at SDSU, Preston worked at Daktronics in the Customer Service Repair Center as a student. He joined Daktronics full-time as a Scoreboard Sales and Service trainee after he graduated from SDSU with an Electrical Engineering degree. He moved to the Seattle office and worked with Marlo Jones for three years.

"In those three years, in Seattle, the office had significant growth in sales from about 300 thousand a year to about 900 thousand a year.

Jim Preston works with his office coordinator in Lenexa, Kansas.

Extra! Extra!

Daktronics Daily Newsletter
Extra_Extra@daktronics.com

Retired Technician "Not Ready to Sit Still"

If you glanced at Mitch Arney's resume, you would notice one thing: he is not your typical service technician. His previous job experiences include: eight years of active duty in the U.S. Navy, 19 years of service in the Naval Reserve, and 30 years with the Iowa State Patrol Communications. Then, in December 2006, after more than 40 years in the work force, Arney decided to slow down and retire.

Mitch Arney (LE Services) has a passion for his work with Daktronics.

To pass the time, Arney wired his sons' houses, helped his daughter finish her basement, installed cupboards in his other daughter's condo, and continued volunteer work with the American Legion.

"I just felt like there was more that I could do ... I was not ready to sit still," said Arney. "I wanted to be out and about, and feel like I could be productive."

So, Arney did what any self-respecting professional does, when they believe they have the opportunity to continue their career - he skipped out on retirement. During the winter months, Arney searched for a job working with electronics, not out of monetary necessity, but for an entirely different reason.

"A lot of people have hobbies when they retire, it just so happens, electronics is one of mine. To me, this is more of an enjoyment, not necessarily work. It's something I truly enjoy."

In February, the 60 year-old was hired as a Service Technician at the Daktronics Sales & Service office in Ankeny, Iowa, where he'll continue fueling his passion and developing his electronic skills.

"To me it's pleasant, it intrigues me, and I'm sincerely interested in not only the product, but also the technology behind it. It's very interesting to me...I take pride in my association with Daktronics."

This was due, in my opinion, to a growth in product offering like MC and LED technology for indoor sports," said Jim.

In August of 1998, he moved to Kansas and opened the Topeka office. Then in 2007, the Daktronics Sales and Service office moved to the suburbs of Kansas City, in a town called Lenexa. The current staff includes three retired athletic directors, two techs, and a commercial salesman.

"When I moved to Kansas, there had been a top competitor reseller in the state for the past 25 years. I found one Daktronics customer near Topeka from about 15 years before and three or four scattered around the

Kansas City Royals – World's largest HD display measuring 105 feet high and 85 feet wide.

state. The one thing that I was doing that they were not, was physically visiting the outlying areas and providing on-site service. As the word got around about our on-site service and working on competitor's product I got more phone calls. We got a small project at Kansas University and I started referencing that in my sales presentations. We now have major installs at three of the major Universities in the state; Kansas State University, Kansas University, Wichita State University, and the Kansas City Royals installed the largest HD display in the world in April 2008. These large projects and the service we give ads a lot of credibility to Daktronics. The biggest High School Parks and Recreation project I've ever done was at the Topeka Sports Park (home to three high schools). Three separate high schools wanted a single facility to hold events for all schools. We installed football, baseball, and softball products. It was the first conglomeration where multiple school districts used the same field."

Topeka Sports Park

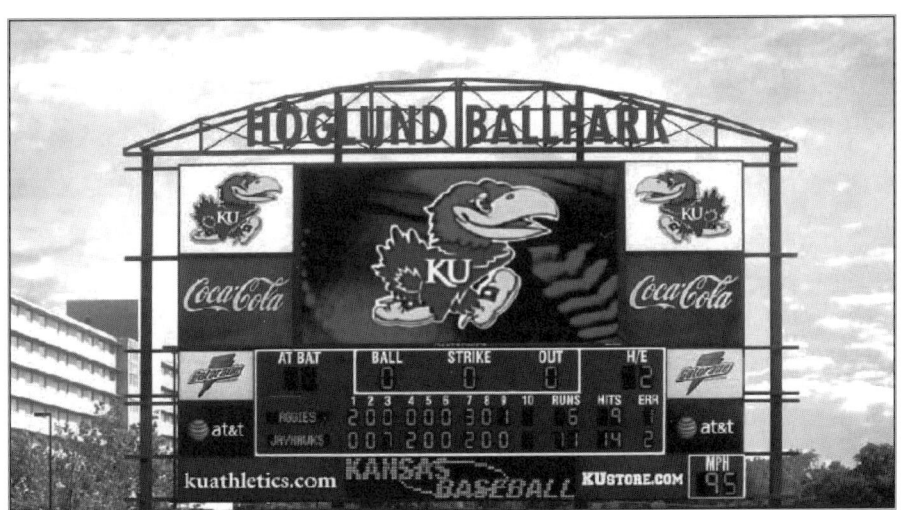

University of Kansas Hoglund Ballpark – one of the many displays purchased by KU.

Retired Athletic Directors Bring in Business

Each year Jim Preston sets the goal of $1 million for the Lenexa, Kansas, office. Last year they reached $950,000 and are expecting to reach the $ 1 million goal in another year. He attributes a good portion of the office's success to the retired athletic directors who have been hired throughout the years. Their contacts and ability to get into schools is matched by none. Steve Grant talks about how working at Daktronics has changed his professional life.

Steve Grant

"I was an educator in the public schools in Kansas for 35 years. Working in my new career with Daktronics has represented quite a change. While I miss the day to day contact with students, coaches, and teachers, I have been able to visit quite a few schools and interact with athletic directors who used to be my colleagues. I have accompanied a couple of retired A.D.s---Dick Schumacher and Art Newcomer on several sales calls. They have been a tremendous help. I have really enjoyed the opportunity to work with high school athletic directors and make recommendations and suggestions on various products that will enhance their athletic programs. My biggest job to date, has been the sale of a football scoreboard with ad panels and delay of game clocks to Paola High School in Paola, Kansas. This was very satisfying...being able to work with the athletic director to determine the needs of his school."

MINNESOTA

"Diverse Customer Base" Drives Sales for Eden Prairie Office

Adam Skare, Tom Siemonsma, and Joey Hulsebus opened the Eden Prairie, Minnesota, Daktronics Sales and Service Office in the late fall of 2006. Adam Skare says that "the customer is the number one priority. Making this a priority guides our product sales and increases capacity for excellent service in the future." But, before providing excellent service, the

Joey Hulsebus   Tom Siemonsma   Adam Skare

Eden Prairie office had to be moved into. Although teamwork is one of many values at Daktronics, sometimes it comes down to one man getting an office situated and in working condition.

"I remember the first two months the office was open. We literally had 14 chairs and a round table. When the long awaited day came for our office furniture to arrive, I was the only one there and the cargo truck driver asked me where my fork-lift was to unload 20 or so crates of furniture. But since we obviously didn't have one, I had to unload the truck by hand," said Adam Skare.

Once an office is properly organized it can start reaching out to potential customers. Starting the Eden Prairie office in one of the oldest Daktronics markets was not simple. As Adam points out, it's important to bring a fresh image of Daktronics to the area when building and rebuilding relationships.

"Our philosophy on selling may be slightly different than other offices as we have a very diverse customer base, most of which have previous Daktronics experiences, both good and bad. Our sales philosophy is to continue to support valued customers as well as try to re-engage those customers who have had negative experiences with us in the past," said Adam. "Our office is bringing a new level of service to the Minneapolis area. Being one of the oldest markets for Daktronics made it a challenge to come in and try to build a new image. Virtually every reseller in Minnesota knows Daktronics for one reason or another, but it was our challenge to show them the new Daktronics and services we can offer."

High Profile Signs Help Office Reach Goals

The greater Minneapolis area has the potential for developing a large customer base. According to Adam Skare, Daktronics displays in high profile locations contribute to gaining business.

"Financial goals aside we've had great progress in penetrating the downtown Minneapolis and St. Paul Markets. We've had several opportunities to bid or participate in some very high profile projects in the metro area. It's always great to get projects, but when they are showcased in an elite location, that's really special," said Adam Skare. "One of the first commercial projects in the region was at Mac Daddy's in St. Cloud. It was a truck stop that wanted to really make an impression so they purchased four Galaxy® Pro 20mm units, plus some digits. It was a real spectacle for

Mac Daddy's Truck Stop

the area and many new leads came from the project."

Other projects that the Eden Prairie office is particularly proud to be involved with include: Becker Furniture World, Elk River Ford, Jackpot Junction, and Westminster Presbyterian in downtown Minneapolis. "That is a very high profile intersection in town and the project took over a year to close, but the expose will be tremendous," said Adam.

Jackpot Junction

'It's working!'

Gas stations are the hub of everyday life and are full of people rushing from one place to the next. Providing service at gas stations can turn into a disaster; especially if the service interrupts regular business. Telly Masse talks about one time he accidently put a gas station out of commission for a few moments.

"In February 2007, I was performing the in-store part of the installation of a Galaxy® display at a gas station in St. Bonifacius. To complete this install, I needed to connect a Cat5 cable to the

Telly Masse

switch located in the cabinet under the register. Shortly after I proceeded to pull the rack out, I noticed a Cat5 cable coming out of its slot (the connector was obviously loose). I plugged it back in immediately and was hoping that nothing got interrupted by it. I proceeded to connect my Cat5 cable and put the rack back in its place. After I got done, I stood up to look at the Galaxy® display outside when I noticed about ten people heading towards the store from different pumps. I was praying that it had nothing to do with the Cat5 cable, but as soon as they walked in, they started to complain about the pumps refusing their credit cards. I kneeled back down as fast as

I could and double-checked to make sure that the cable was in all the way. They stood at the register staring at me for a good five minutes. These five minutes seemed like forever because I knew it would just take a little time for the connection to be back up. I kept apologizing to the clerks and the clients. I was so relieved when one of the clerks screamed, 'It's working!' I learned a big lesson from this incident and now I always make sure that all the Cat5 cables are properly connected before I touch the rack. Somebody else's poor wiring job can cause us a lot of grief."

Quite the First Day

Adam Skare learned the hard way to always dress for the occasion. "When Tom and I were first planning to relocate to Minneapolis, Kelly Koenig sent us up to do some demonstrations and get a feel for the job. One of these demonstrations was in Duluth where we were to show the 12mm unit. It was my first real demo so I prepared and put on my nicest clothes thinking I wanted to look great for the new job. Only to find out that our demo was at a coal yard! It had been raining all day and we were ankle deep in what at first appeared to be mud but I eventually realized was coal dust. Needless to say those pants, shoes, and socks all ended up in the garbage. Quite the first day."

There's No Work Like Teamwork

Daktronics functions as a team. Joel Bertram was able to complete an assignment that was not part of his daily duties with the help of the Networks Operations Center (NOC).

"During my second week in the Eden Prairie Daktronics Sales and Service, I was asked to go to Rochester, Minnesota, to help reprogram a modem in one of our billboard displays. Although I have limited technical training, our Field Service Engineer was out of state on an install and I agreed to the assignment. With the help of a rep-

Joel Bertram

resentative of the Daktronics Network Operations Center, I was able to flawlessly complete the assignment. While this was my first foray into field service, it was a success and was possible due to the idea that no matter what has to be done, we need to service our customers. It was the Daktronics winning attitude that allowed me to complete a task that was not in the realm of my normal duties as an Office Administrator."

Missouri

Hard Work Pays Off in Earth City

Bryan Nagel and Ken Green opened the Earth City, Missouri, Daktronics Sales and Service office in August of 2004. Ken had 30 years of experience in athletics before his work started in the St. Louis office. Bryan moved from the east coast and looked forward to the opportunity of opening another office.

"My family has also been very happy moving back to the mid-west. My wife Teresa and I are both from South Dakota, we lived in Baltimore for 12 years and she really missed the Midwest culture. We have three children, ages 8, 10, and 15. They all enjoy Missouri and I have enjoyed the opportunity to yet again start off a Daktronics office and all the joy of seeing the hard work pay off."

Ken Green

Hard work was right. Bryan and Ken made as many as 50 calls a week; driving from school to school dropping off cards and catalogs. Nathan Hendin joined the sales team as a Retired Athletic Director and brought his wealth of contacts to the office and helped put Daktronics on the map.

"Since I knew everybody and Bryan knew everything, we set out to visit every school in our

Bryan Nagel

Busch Stadium

region. Very few people knew about Daktronics scoreboards; clearly we had our work cut out for us. We had to fight tooth and nail for every sale and make our presence known in each opportunity that came about," said Nathan. "Every sale, whether big or small, was treated as if we were selling equipment to a new major league stadium."

A few months later they had a firm contract on Busch Stadium and were able to use the renderings of the stadium as a lead in for sales. "One of our first breakthroughs was Farmington High School. They were looking for a standard football scoreboard. With the help of our RGB demo trailer and the fact that they had 26 sponsors for the stadium, we were able to put in the first full-color message center (MC) in the high school. Since this opening we have installed 15 football scoreboards with full-color MCs in the last three years. Another example of our success was Hermann High School. This school was a strong competitor's school, but after three or four meetings and bringing in Track timing with Finish Lynx, we were able to put in a large football/track system with scoreboard, message center and first Sportsound® 1000," said Bryan.

Farmington High School   Hermann High School

Chaifetz Arena at St. Louis University

Scottrade® Center – home of the St. Louis Blues.

Starting in a new area with only a handful of Daktronics' boards, Bryan and Ken gladly serviced competitor's scoreboards. Bryan says that this only gave them an opportunity to start and build relationships with customers and generate business.

"When the office first started we were still servicing all brands of scoreboards. We were able to work on a few boards that built a good enough relationship to win replacements board orders. One example is Pattonville High School. They now have two baseball scoreboards and one football scoreboard from Daktronics."

Opening Doors at Schools

Bryan attributes the success of the Earth City Daktronics Sales and Service office to a dedicated staff and large installations that helped put Daktronics' in the public's eyes. After which Bryan and Ken were able to close sales and expand throughout the area.

"Along with Busch Stadium, we have done new boards at Mizzou Arena and Faurot Field at the University of Missouri in Columbia, all new equipment at the Scottrade® Center (home of the St. Louis Blues) and just now completing the Chaifetz Arena at St. Louis University. All of these large profile facilities really help open doors at schools. I never imagined we would grow at the rate we have. In a matter of two to three years we grew the office to a staff of ten people, and in Fiscal Year 2008 we will have over $ 1.4 million in sales. This surpassed my wildest imagination. I feel that having three retired athletic directors in our office makes a large difference in the growing of the St. Louis market. We have a great staff and they do a great job getting doors opened. Tim Moore has done a great job covering southern Illinois. He is in the enemy's backyard, with the competitor's headquarters being only fifty miles from his house. He has done a great job convincing

Alton High School

customers to give us a shot. We now have a large number of Daktronics boards covering this area. We must be doing something correctly; I received a job offer from the sales manager at a competitor's office after we beat them out of a large scoreboard project at Alton High School."

Commitment to Service Keeps Republic, Missouri, Office Successful

Since the Republic, Missouri, Daktronics Sales and Service office opened its doors for business in 1991, numerous sales and service goals have been fulfilled.

In the beginning, the Republic office was headed by Dave Marsh. Not long after, Joseph Hicks joined the team. He eventually turned the office into what it is today; a one man office that has reached service goals above and beyond the competitors. The one-man Republic office has even managed more sales compared to other office that have a larger staffs. In 2007, Hicks sold over $700,000 in Daktronics product while taking care of all the daily responsibilities in the office that many of us overlook.

Joe Hicks

"I am a one man office. I sweep, I clean, I sell... whatever you need, I'm your man!" says Hicks.

Some of the projects that Hicks is particularly proud to have sold and serviced are local high schools in Missouri. These include: Harrison High School, Ozark High School, and Hollister High School which is particularly proud to be a "Daktronics Only" high school, meaning Daktronics is the high school's top choice when it comes to its LED needs.

Commitment to service is just one of the many reasons why the Republic office is so successful. When asked about his philosophy on selling and servicing, Hicks stated, "My philosophy on selling is more of an action than a phrase. I believe that being in front of the customers, talking to them, stopping in to say hi, sitting and chatting with them to see how

Mountain Grove High School

Ozark High School

they are doing, and servicing the product in a timely manner is what makes a person successful. As a salesman, you have to be willing to be available, and that might mean staying up late or getting up early."

One specific example of Hicks' philosophy comes from an experience at Ozark High School in Ozark, Missouri.

Steve Brashears working in the Republic, Missouri, office.

"Ozark had purchased a colored marquee sign for the front of the school. There was one week that they had no power going to the board, and a special event was going to be happening. Every morning before school, I would get a generator out there and continuously fill it with gas throughout the day so that they could display the information. That's probably why they now buy all of our stuff!" said Hicks.

When asked if he had ever experienced any embarrassing moments in the field, Hicks replied, "I was doing a demonstration for Sportsound® equipment in Mountain Grove, Missouri, when all of a sudden the cops showed up. They had come to resolve a noise complaint. I quickly explained that we were doing a demo that would take 30-40 minutes. They were very cooperative and let me finish the demonstration."

Support From Union School District

PO Box 440 Union, MO 63084

December 11, 2007

Dear Daktronics team,

Union High School just completed a nine million dollar gymnasium project that is one of the finest gyms in the St. Louis Metro Area. The crown jewel of the gym is our Daktronics center hung scoreboard.

We have heard nothing but positives from the community about the gym and the scoreboards. We recently held our annual wrestling tournament and having the ability to run three scoreboards for three different mats was awesome.

We also have the full color message center on two sides of the center hung scoreboard; we run the players' rosters, pictures as well as sell advertisements as a fundraiser for our marketing program DECA.

We also purchased a new Daktronics football scoreboard we were very pleased as well. Our team did very well so we had large crowds and we heard many positive comments about the scoreboard.

Bryan has been awesome as far as training and service on both boards. Whenever we have a question he is just a phone call away and has had every situation resolved in a quick matter of time.

I would highly recommend Daktronics to any school or business that is looking for scoreboards or electronic signs. They are a top class company.

Sincerely,

Jeff Van Zee
Director of Facilities and Operations
Union R-XI School District

Customer Comments

"We are extremely pleased with our new scorer's table in the gym. The quality is second to none! The whole process from start to finish was very easy and the service that Daktronics provided was outstanding."

KURT SIMON
Athletic Director
Carmi-White County High School
Carmi, Illinois

"It was a pleasure working with Daktronics. Tim Moore was extremely knowledgeable of his products and very interested in getting us exactly what we needed. The quality of the product was first rate. From beginning to end it was a great experience."

BILL SCHMIDT
Athletic Director
Belleville West High School
Belleville, Illinois

"We are convinced that Daktronics makes the best running and best looking scoreboards on the market. The group at Daktronics is honest, knowledgeable and enjoyable to work with."

JACK MARTIN
Facilities Director
Alton School District #11
Alton, Illinois

"Joe Hicks and Daktronics were great to work with. From the start to the finish the whole project was very good. Our new football scoreboard looks great—everyone is happy with how it turned out."

MIKE WILSON
Athletic Director
Strafford R-VI Public Schools
Strafford, Missouri

Nebraska

Building Relationships in Lincoln

Bob Howell officially opened the Lincoln, Nebraska, Daktronics Sales and Service office in September 2005. Before that, he was an independent, Daktronics dealer for S&W Equipment; which sold Daktronics products since 1984. S&W Equipment was one of the first resellers to work with Daktronics. Bob purchased the business in 1995 and ran it until 2005, when it was changed to a Daktronics Sales and Service office. The Lincoln office has been growing Daktronics' business by exceeding their goals ever since.

"The first year of business we did a little over $300,000 and this fiscal year we will reach the $1,000,000 level in the High School Park and Recreation market. Our service goals are to respond to customers within a 24 hour period and we're able to meet that every year. Some featured projects include: Skutt Catholic High School, in Omaha, installed a BA-3718 baseball board, and since then, Daktronics has replaced all of their competitor's equipment. We have also added a Galaxy® message center at this campus. In Adams Central High School in Hastings, Nebraska, because of our relationship with them, they replaced a competitor's board with our equipment. We also sold them a message display. They were so pleased that they called us to replace their football board. It's probably all due to our service and customer relations. We're always on call and accessible to our customer for their service needs...this builds relationships. We strive to make the customers feel that they are #1 to us."

Part of building relationships is making people feel as

Bob Howell

Adams Central High School

if they are the most important customers. An example of this would be our willingness to help a school district when their signs had been mangled from a deadly storm.

"Norris High School, in Firth, Nebraska, had a devastating tornado that came through and damaged all of their buildings; the high school, middle school, elementary school, bus garage, and press box/concession Stand. We replaced their football scoreboard that had a 15 passenger van picked up and thrown through the scoreboard by the tornado. In addition to the new football scoreboard, we replaced their softball board, the middle school boards (one of which was never found) and serviced/repaired the high school and elementary scoreboards. We were able to get all of the boards up and working right before the season and met their tight deadline."

Burke High School

The remains of the Norris High School football board (center left) after a 15 passenger van (center right) was thrown through it by a tornado.

Simple Solutions in Tight Circumstances

Since Bob has been in the business for quite some time, he understands that sometimes angry customers require simple handling.

"We got a call from a customer with incandescent boards and one board would not light up. I asked if it was plugged in...they were quite irritated that I even asked that. They were pretty mad that it wouldn't light up and it didn't help that they had a game that evening. After an hour drive I arrived there to determine what the problem was. I got on a ladder, went to the top of the board, "plugged" in the scoreboard, and it was up and running for the game. Apparently a student had knocked the plug out of the wall with a ball of some sort."

When things go wrong during public events, this can add a whole new level of stress. Another simple solution involved the operator not knowing the difference between basketball and volleyball mode.

"Another High School was quite upset when they started their basketball season. At the end of the first quarter, they hit the second quarter and the scoreboard's score went blank. I determined they were in volleyball mode not in basketball mode. I responded in a timely manner to get them going."

Daktronics Installs Six ProStar® Boards in One Day

Tim Nichols, a Field Service Engineer, has worked at Daktronics for 10 years; 3 ½ years with Sports Products Engineering, New York for 1 year as a Daktronics Sales and Service manager, and in Wisconsin for 5 ½ years as a Field Service Engineer. He hit a milestone in his career during the summer of 2007, where Daktronics and Lamar teamed up in Janesville, Wisconsin, to install six Galaxy® displays in one day.

"I spent a couple of days before hand to get the control cabinets working with the modems and routers; making sure the NOC was able to control the units. The following day we met at the Lamar office in Janesville, where we built all six displays and had them secured to two large trailers; three displays per trailer. It was on Saturday morning at 4 a.m. when we were all met at the shop for our morning sugar rush of coffee and doughnuts," Nichols continued. "Once we broke into our teams, we headed out to our first site which included people from Daktronics, Lamar, and Westphal (our ASC). Once the first board was up, secured, and working with our test patterns; we moved on to the next site to begin the process all over again. By 2:30 in the afternoon, all boards were up and running with default content. It's all in the customer's perception. First, we knew coming into this task what was expected of us and we all were focused 110 percent on what was happening. From installing the control units, assembling the displays, and teaming up to get these all raised and operational in one day, was a combined effort from all companies coming together to accomplish one task; make our customer happy. Perception is reality!"

Tim Nichols

Tim Nichols contributes the success of the installations to what he calls the "A Team." This includes: Alan Hess, Tony Spetz, Chris Parker, and Mark Stopple of Daktronics as well as team from our authorized service company; Marty Kelly, Brent Folmer, Rick Emerson, Steve Mewis, and Jeremy Bierman for keeping focused throughout the entire process and not once looking back.

"This project is proof that when you put your mind to it, you can do anything. I would have to admit this project was very impressive considering it usually takes about 16 hours, on average to put a display up. The Daktronics' team did a lot of preparation to make the installations a success. For all of Daktronics' work, the boards were installed and content was put on the boards the same day. Lamar was very pleased with the installation," said Nichols.

> "I just wanted to let you know that our digital network installation this past Saturday went very well. We had a lot of coordination going on between different parties and it had the potential of being very confusing. In fact, we had two installation crews working simultaneously starting at 4:00 a.m.; we had all six faces hung by 2:00 p.m. and all were operational by 6:00 p.m.

Lamar Billboard

Daktronics sent in the troops (Alan, Tim and Tony) to make sure the installation went smoothly. Those guys worked VERY hard to make sure we had a successful install. In fact, Tim Nichols continued working until 10 p.m. that night, riding the market with me to ensure that the displays were not burning too brightly.

Please pass along my compliments to the folks at Daktronics. Their staff deserves major kudos!

Brad Yarmark
General Manager
Lamar Advertising of Madison – Rockford

The Ballerina Act

"Probably the most pervasive quality we have that we need to practice on a day to day basis is to be helpful ... We have to be helpful to fellow employees. Anybody can make mistakes and everybody will make mistakes now and then, and we have to understand that and we need to be helpful. We have to help them through their difficulties and help them do a good job. Help them learn more, help them perform better because that reflects back on our team," said Dr. Al Kurtenbach.

Every person finds themselves in a slippery situation beyond their own control; Tim Nichols is no different. While he was finishing an installation at Cornell University in New York at their hockey facility, Tim found first hand that Daktronics' employees go to great lengths to help each other through the iciest of conditions.

"I was finishing up new locker room clocks that needed to have a wire installed to allow the newer clocks to run with the new controllers. I was walking towards the water fountain to wet my sponge when my foot slipped out from under me and ripped my ham string so severely I could hardly walk. Murphy's Law was with me that day. I was 3-1/2 hours away from home and the muscle I ripped was my right leg, the one I used to control my vehicle. Just a couple of hours earlier that day, I found out that my best friend from Daktronics, Dave Dininger, was in New Jersey and wanted to come to our house and visit. We were supposed to meet for dinner that night and I was just about finished and on my way home when I decided to learn to dance on the ice. I ended up calling my wife and then Dave to see if they would come out and pick me up. My wife would bring Dave out to my location and he would drive the Daktronics service van while my wife trailed behind us on the longest 3-1/2 hour drive of my life. Without him being there, I don't know what I would've done. After that, I had to work at home, making sales calls for two weeks. It wasn't until June of that year did I get coined the "Ballerina Dancer" from Dave. Sitting down having lunch around several people at the Swiftel Center in Brookings, Dave would grab everyone's attention and then pop the big question: "Hey did you guys hear about Tim's ballerina act?" Dave would start laughing while his eyes pierced mine knowing the truth was about to tarnish my dancing career. Truly, this was an experience I would never wish on my worst enemy. I consider myself very blessed knowing my wife and my best friend Dave was there for me when I truly needed them."

South Central Region

Arkansas

A Foot in the Door in Little Rock

The Little Rock, Arkansas, Daktronics Sales and Service office was opened in July of 2005. The office evolved from a commercial salesman-in-a-van (Jim Vasgaard) in Jonesboro, Arkansas, to an actual fully staffed office in central Arkansas. Part of being a new name in the area is establishing the company as a viable option for customers who are in the market for a new display. Valorie Stoll, office coordinator, talks about how the office is finally breaking into the territory as a serious contender.

"A strong regional dealer of a nationwide manufacture, based in Little Rock, Arkansas, has quite a monopoly on the scoreboard market in Arkansas, and in all honesty, most of the mid-south including Mississippi, Louisiana, and Texas. The introduction of our Daktronics Sales and Service office into the area has broken into that monopoly and brought product and pricing benefits to customers. For years, schools and park departments just took whatever board they were quoted and didn't really have the opportunity to discuss options. Even after two and a half years in this state we still hear, "I had no idea that was an option or that we could even do that." Schools are hungry for unique design because for so many years they've only been offered a cookie-cutter set-up that all the other schools get. This dealer has one look for boards without message centers and one look for boards with message centers. Our niche is the school marquee - that gets our foot in the

Jim Vasgaard

Valorie Stoll

door. We have great dialogs with schools that are considering message centers with their new scoreboards. We're constantly looking for new ways to accentuate Daktronics positives, without mentioning competitor weaknesses, and close more deals. Having company employees to provide service is a strong selling point to customers outside central Arkansas. Those areas see additional companies marketing to them, but they just don't have the service options that we do. Right now we're still in the phase of developing name recognition and market share but we see improvement every month."

Acts of God

The Arkansas office has received a couple of big orders in high schools throughout the area. Some orders are a result of natural disasters. As Bryan Morris, lead tech, explains, the last standing company usually wins.

"Recently, we installed a marquee at Mountain View High School. A week and a half later, a tornado swept through the area wiping out most of the town including the school's football board, but our marquee was left standing. The force of the storm pulled the face of the 3200 series Galaxy® open, broke the hinges, and pulled ribbon cables out of the latches. Once power was reestablished to that town, it turns out that

Bryan Morris

The Mountain View High School marquee left standing amidst the tornado damage.

Demo trailer

not only was the sign left standing, but it still worked. The assistant superintendent we've been working with is happy with our relationship and our response to the school's needs. We've provided information and pricing to the insurance company and have spoken to the adjuster a couple of times to explain the particulars of the equipment. It isn't every day an adjuster has to work a claim on a scoreboard."

Repeat Orders

Daktronics Sales and Service offices are dedicated to meet the needs of customers. Bryan Morris, lead tech, says that "remaining flexible, finding creative solutions to the infrequent problem, and understanding the school's needs, are the keys to getting repeat orders. Things may not always go perfectly on a school's project, but the staff is diligent in keeping their customers satisfied with both the company's products and the office's service. We expect long term benefits from our actions, but the short term pay off has been greater than anticipated. Positive references from the customers that had less than smooth installations have really added to our customer base. They see how hard we work to complete their project."

Morris mentions a creative solution to longer lead times. "In 2006, the plant lead times were extended to 18 weeks. In order to satisfy a school that ordered a football board, with a message center, the office took the demo trailer to the games to keep the customer happy. Now, because of the relationship, we're in talks for a new basketball board"

LOUISIANA

Baton Rouge

In 2001, Mike Durston re-opened the once closed Baton Rouge, Louisiana, Daktronics Sales and Service office in a small rental unit in which he was the sole employee. Today, the six office members have set a goal of reaching the One Million Dollar Office, and has come close to reaching their goal. In order to work towards reaching their goal, the office and Durston has had some interesting and memorable experiences.

The first scoreboard that Durston ever sold was to Airline High School in Bossier City, Louisiana, in 2001. After the installation, Durston took a photo, sent it in to Daktronics' corporate office and a marketing department put it in a catalog. Seven years later, Durston was meeting with the principal of Airline High School and the principal told Durston that while attending a conference, he saw the photo of his school at the conference. Both Durston and the prin-

Mike Durston in the Baton Rouge Daktronics Sales and Service storage area (2001).

Airline High School

cipal were surprised at the coincidence.

After Hurricane Katrina nearly demolished the city of New Orleans, Durston and his office stepped in to help rebuild the city. After the disaster, Durston had three families staying in his house for over a week without amenities and electricity and in 102 degree temperatures. The pressure did not subside at work either, the Baton Rouge office was working at the Louisiana State University stadium in order to update the control system to NFL standards. Since Hurricane Katrina ruined the Sugar Bowl in New Orleans, the university was going to be hosting several of the NFL games and that required Durston and his team to put in extra time and effort to meet the rigid league guidelines and deadlines.

Baton Rouge service van.

Mississippi

Hattiesburg – On the Edge

Mark Carr opened the Hattiesburg office in 2006 and is starting to get things rolling in the market share. Carr describes the Hattiesburg office as one that deals mostly with smaller boards, in the High School Park and Recreation market, with a lot of projects "in the works."

"Our competitors have a good reputation and market share. We've been winning battles here or there, but it takes a couple of years for a salesperson, or sales force, to generate revenue and start to see a return on an investment," said Carr. "We're on that edge where Daktronics will start seeing a return on its investment. This year will hopefully be the break out year."

Projects that the Hattiesburg Daktronics Sales and Service office has worked on include almost all of the University of Southern Mississippi's sports facilities. Columbia High School is also a notable project as it was the office's first large football board project. The office was able to win the contract against the competitor. This was the same school that Walter Payton attended, so a bronze statue was constructed, under the scoreboard, in his honor.

Thick-Skin

Persistency is key when breaking into a new market. Mark Carr learned that having "thick-skin" is important when working with difficult people.

"I had a customer (a bottler) who became quite upset when I started visiting the schools in the bottler's sales territory. He told me that if we continued to visit his customers, he'd make sure that we wouldn't sell any boards in the state of

Mark Carr in his office in Hattiesburg (2008).

Columbia High School scoreboard – the same high school Walter Payton attended (notice the bronze statue under the board).

May 5, 2008

201 Daktronics Drive
Brookings, South Dakota 57006

Dear Daktronics Team:

We opened the Larry Doleac Baseball Complex this spring for the Hattiesburg Dixie Youth Baseball League. The complex includes five fields with five scoreboards from Daktronics. We are extremely pleased with the quality of the scoreboards and service that we have received from the Daktronics office here in Hattiesburg, Ms. A spokesperson with major League Baseball's "Baseball Tomorrow Fund" stated that our park is "one of the best he has seen in the U.S." and Daktronics was a part of that recognition.

Mark Carr has delivered on everything we needed plus some on this project. He was very hands-on, meeting any deadlines that we had. Terra George was also a great resource in helping with the design of the scoreboard and helping us get the correct logos on the sponsor panels. I appreciate working with your team in Mississippi.

I highly recommend Daktronics and would be happy to work with them in the future. The concept of Excellent Customer Service is alive and well here in Hattiesburg, Ms. Again, thanks to Mark, Terra, and your service team. They did an excellent job.

Sincerely,

Freddie Triplett
Hattiesburg Dixie Youth Baseball

Mississippi. He didn't want new pressure, he wanted to choose who and when. But we've made head way and have recently sold a 28-foot baseball board, two sets of basketball boards and hope to make some football product sales soon. You just have to have a thick-skin and keep going back to the customer and deliver when you have the opportunity."

OKLAHOMA

Taking Daktronics to the Next Level in Norman

When Perry Grave started working for the Daktronics Sales and Service office in Norman, Oklahoma, he was the sole employee running a "one man show." He was responsible for the sales, service, installations and everything in between. Before Grave started working at Daktronics, the office had been a "one woman show." Lisa Glanzer, the daughter of co-founder Al Kurtenbach, started the office in her Norman, Oklahoma, home in April 1994.

Today the office has eight employees and has moved from its single occupancy office space to a double occupancy with 3000 square feet. In spite of the relatively small nature of the office, it has attained large goals; it was the first Daktronics Sales and Service office to reach $2 million in sales.

Perry Grave in the Norman, Oklahoma, Daktronics Sales and Service office.

Even though Grave did not sell the displays at the University of Oklahoma, he cited this project as the one he is most proud of. The University of Oklahoma had installed the first generation of ProStar® displays. Grave was invited to be a part of the pre-install meetings to show the local service presence. He said, "I was a part of something that helped launch Daktronics to the next level."

The philosophy for the Norman, Oklahoma, office is to put the customer first and keep their best interest in mind when selling and servicing.

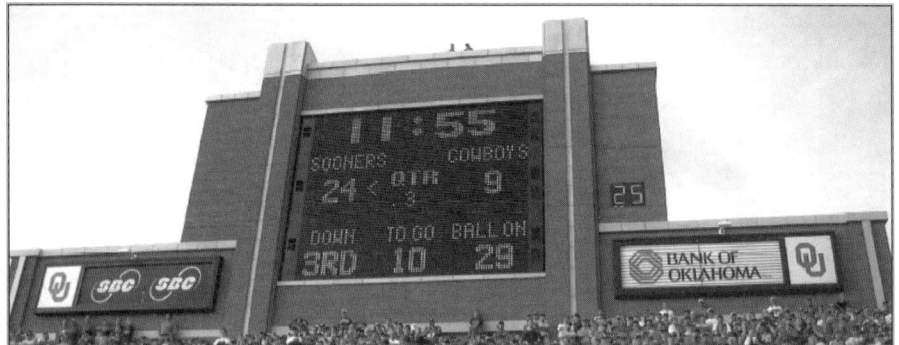

University of Oklahoma

"If it weren't for the customers, we wouldn't have jobs," said Grave. "The customers are the reason that Daktronics is what it is."

Not everything with Grave's career has been something he is proud of. During a sales conference presentation, Grave flipped on a shot clock he had displayed behind him. After a few minutes, he noticed a panicked look from some of the people in the front row. Grave turned around to face the shot clock leaping with flames. After the presentation, he discovered that the display had been loaded with the wrong type of resistors. Even though his presentation was a bit of a mishap, it was still relatively successful.

"I still got three sales out of it," said Grave.

Texas

Delivering in Dallas, Texas

The Dallas, Texas, Daktronics Sales and Service office was opened in 1996 by Michael Howell. Even though Howell officially came on board with Daktronics in 1996, he had been working extensively with Daktronics products since 1984 with National Display Systems. National Display Systems was the exclusive Daktron-

Mike Howell in the Dallas office

ics representatives in the Texas, Oklahoma, Louisiana, Arkansas, Mississippi, and Tennessee areas until Daktronics bought the company in 1996. After Daktronics acquired the company, they opened the Dallas Daktronics Sales and Service office and brought Howell on board to get things started.

Waco Independent School District

The Dallas Daktronics Sales and Service office is Daktronics' largest Daktronics Sales and Service office with around 41 employees. To accommodate the growth, the office has never had to actually change addresses, but has expanded the building three separate times during the 12 years it has been open. One of Howell's most memorable sales for the Dallas office was the sale of the Waco ISD Sports Complex displays in Texas. The Dallas team secured the first ProStar® sale and installation by a Daktronics Sales and Service office in the high school market, which was a very significant sale at the time.

"That was a lot of fun. There was a lot of competition to try and get that one closed down. That was very memorable for me. It was kind of a source of pride to be the first office to sell to a ProStar® to a high school market."

Shortly after this sale, the office was able to sell two more ProStar® displays to Mesquite High School, which is right outside of Dallas, Texas. Those two displays alone constituted a sale of about $1.7 million.

Throughout the years, the Dallas office has been very successful in reaching their goals. They continue to effectively sell and install boards throughout Texas and the surrounding areas. Part of the office's success could be contributed to its sales philosophy—treat the customer as you would want to be treated.

According to Howell, Daktronics employees don't just meet expectations, but they try to exceed expectations. This sales and service philosophy is also applied to the great support Daktronics provides after the initial sale and installation. The company has been able to build many strong relationships based on our well-known service and concern for the customers.

Texas Rangers

"There's the pride and accomplishment of actually making the sale, but another part of it is the fact that we're pretty unique in building strong relationships because we have such a good product and good service after the fact too," Howell said about the philosophy.

Ed Weninger teaches Mike Howell how to use Maximizer.

For the past 24 years, Howell has had many unique opportunities and experiences through Daktronics. One of his most memorable experiences goes back to servicing the Texas Rangers displays. The Rangers had just installed a speed pitch indicator, but they were having trouble getting it fired up for the first game of the season. Lucky that he was on site, Howell took off down the hall in the stadium to try to find the correct circuit card to fix the problem.

"There was a guy standing in the hall who was wearing his cowboy boots and they happened to be sticking out a little in middle of the hallway. As I was running by I kind of kicked one of his boots. The two guys who were with him looked at me funny, and that was when I realized they were the Secret Service guys, and I had actually kicked the future president, George W. Bush."

Howell credits some of the success of his office to the Retired Athletic Directors program. Howell has been able to hire and work with three former athletic directors in the area. According to Howell, he believes it is a pleasure to be able to work with the former athletic directors because of the ties and experience they have from working in the industry for such a long time.

Overall, the Dallas office has enjoyed 12 successful years with many more to come. As for Howell, he would not want to be anywhere else.

"My job is one of the best jobs in the world. I truly think that because the people I get to work with on a daily basis are truly excited about seeing the products come together on-site. It's really fun to be able to deliver a product like that. It makes coming into the office fun. Every person should have that."

Houston, Texas, Achieving Sales Goals

Joel Heine started the Houston, Texas, office in December 2002 with another Daktronics employee. Now, there are 12 employees based out of the Houston office. The 12 employees form a tight group that works together to achieve their sales goals.

"We pride ourselves in achieving our sales goals each year, in every market," said Heine. In addition to accomplishing the sales goals, the service aspect of the office is also flourishing; Toyota Center and Minute Maid Park consistently sign up for service contracts each year. The achievements in sales and service can be attributed to the team work of the office and the office's service and selling statement, "Working harder than the competition for everything we do."

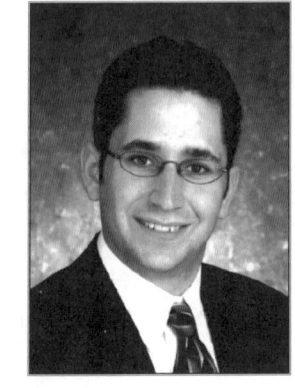
Joel Heine

The project that Heine finds the most pride in is the Navasota Independent School District. "Every person in the office was responsible for a portion of the project. We are team oriented and we respect each other."

The relationship that the Houston office built with Navasota ISD has fostered continued positive service interactions. "The customer is always happy to call us. They are not discouraged from calling us to bring us up to speed—they have a positive attitude and good relationships with us."

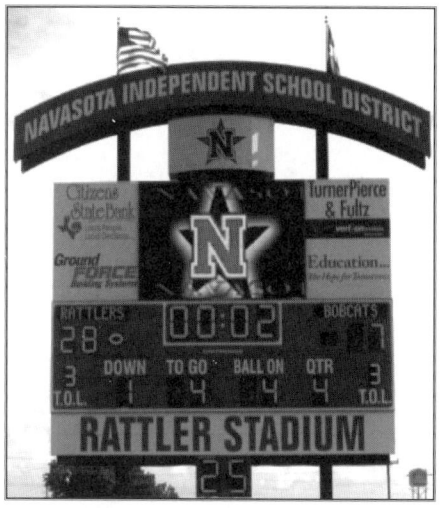

Navasota Independent School District

While working on a different project, Heine experienced something that still brings embarrassment to his voice. He was on a ladder working on a board while talking to Faye Dahl on the phone. He fell off the ladder, but Dahl did not miss a beat. "Faye didn't bat an eye and continued on with the conversation," said Heine.

Falling off a ladder does not compare to another experience Heine had with Frank Kurtenbach. Soon after the Houston office opened, Kurtenbach was visiting the office for a week and decided to take Heine out on a "field trip" of sorts. While the two were driving around Houston, Kurtenbach asked where the Spectrum office was located. At the time, Spectrum was Daktronics' biggest competitor in the Houston area. Heine told Kurtenbach the location of Spectrum only for Kurtenbach to ask to see the office. He agreed and took Kurtenbach to Spectrum.

Once they arrived at Spectrum, Heine expected their visit to be over, but Kurtenbach got out of the car, walked into the office, handed his business card to the receptionist and asked to see the CEO. The receptionist looked at the business card, and obviously startled, ushered the two "outsiders" to the CEO's office. Heine recalls that Kurtenbach and the CEO of Spectrum talked about everything—everything except business. As they left, the CEO agreed to visit the Daktronics corporate office—something he has yet to do.

Expanding Through West Texas – Lubbock, Texas

David Naasz opened the Lubbock, Texas, office in June 2003, in the basement of his house. The office has expanded to include a rental unit with office and drive-in warehouse and has quadrupled its employees to four people. Together, the four employees sell and service over 80,000 square miles and have worked on highly visible projects such as installs at Texas Tech University and Odessa College.

But the office didn't start off with servicing and selling to these well known projects. According to Naasz, at the time the office opened, there were no Daktronics displays in the entire west Texas area. He used some unconventional tactics to make Daktronics more visible in west Texas. The office started by servicing competitor's displays and products and proved that Daktronics could and would provide the customer service that clients needed. It was a way of getting his foot in the selling and servicing door. Today, putting the customer first is still the maxim that the Lubbock office sells and services by.

David Naasz

One of the projects that Naasz is most proud of being a part of is the stadium installation at Texas Tech University. Naasz and his team completely gutted the double T, 1980s display and revamped it with new, high tech LEDs, controllers, auxiliary boards and cabling. In order to ensure the displays were fully functional in time for game day, they worked the

Texas Tech University

entire week and into the night. Their perseverance paid off, the customers were highly impressed with their efforts.

San Antonio, Texas

In the beginning of the Daktronics Sales and Service concept, the offices were able (and promoted) to service all makes and models of scoreboards. It didn't matter who made the board, who sold the board or who bought the board, the Daktronics Sales and Service offices were willing to help anybody with a scoreboard in the hopes of getting the Daktronics foot in the door with the exceptional service that each office offered.

Paul Wildeman in his office (2008).

During Paul Wildeman's first year of starting the Daktronics Sales and Service office in San Antonio in 1994, he serviced various scoreboard models but dedicated most of his time to Spectrum boards. "The feedback I'd gotten was that Spectrum had grown complacent over the years and many of their customers felt taken for granted," said Wildeman. "I became determined never to allow Daktronics customers to feel that way about our company."

Storage in San Antonio (1999).

But becoming competent in servicing the Spectrum boards was no easy task for Wildeman. Everything he

San Antonio parts storage (2008).

did with a Spectrum board was trial and error. "I was bewildered to find their electronics so different from the Daktronics electronics I'd gotten to know," said Wildeman. He relied upon the ingenuity and resources of people at Daktronics such as Brett Wendler and Perry Grave. He also relied on his own creativity and resourcefulness to figure out the Spectrum boards.

"I made a few deals. I sold some scoreboards. And every time I took an old scoreboard down, I traded a portion of my installation fee for the parts out of the old [Spectrum] scoreboard. Over time my parts inventory grew. After a while I no longer offered money, I just offered to haul the old scoreboard carcass away for free! Every time I got some new parts, I'd experiment with them... At times I felt like some kind of mad scientist, working alone, sometimes at strange hours, using my soldering iron as a scalpel, attempting to bring life to lifeless masses."

Sometimes his experiments would work, and other times they wouldn't. One experience in particular reminds Wildeman of his trial and *error* efforts. Shortly after repairing a football clock for a high school with old Spectrum parts, he received a telephone call from the school. They informed him that his "Frankenstien-esque" clock had not worked. They were not able to set the clock over 9:59, which was a major issue since the football level ran at 10:00 time segments.

"I hadn't thought to look for a 10s digit and assumed that all clock packs were the same," said Wildeman. "I had to make a long, embarrassing trip to get back to [the] customer to ensure that they could set their football scoreboard to 12:00 for their varsity game that night. When I got there and rectified the situation, I didn't get much praise. Unfortunately the customer had called Spectrum, the scoreboard's original manufacturer, to find out what could be done. Naturally, our competitor had no kind words to say about me and warned them never to use me again."

Luckily, Wildeman did not give up on his efforts and along the way, he had successes that helped bolster his confidence and also helped him reach the goal of creating a high level of service for the Daktronics name. Each potential customer that he met he would tell them that if they were not happy with their current scoreboard service, that he would like the opportunity to show them what Daktronics could do.

During the winter of 1995, he had a "bite" from a school in southwestern San Antonio. While unpacking his ladder and toolbox at the school, he saw the service man from Spectrum approaching. Wildeman had been caught in the act. The Spectrum employee asked the coach, who had called Wildeman to service the board, and demanded to know what was going on.

The coach quickly came to Wildeman's defense and explained that he had called Spectrum weeks ago and had not heard from or seen Spectrum since. The coach gestured towards Wildeman and said, "I called him yesterday and here he is."

As Wildeman's reputation grew as a respectable and knowledgeable serviceman, Spectrum started warning all of the area schools to not enlist Wildeman for his service. Although some would be disheartened, the Spectrum rebuttal only encouraged him to continue his service and hard work ethic.

"In the beginning, I set a goal to visit every high school, middle school, junior college, and park and recreation organization, in the area," said Wildeman. "But once my reputation started to grow, the phone started ringing and my goal became impossible to accomplish. I soon became reactive rather than proactive and I realized that in order for the business to grow to its potential, I needed help."

He especially realized his need for help when in 1996 he won two contracts for high school football boards—at the time, they would be the largest high school football scoreboards that a Daktronics Sales and Service office had ever sold.

Wildeman recalls the experience of the installation.

"I honed my skills and prepared myself for these two projects that were scheduled for installation in the same week—just days before each school's first game of the season. I had to oversee two different subcontractors, working in cities roughly 45 minutes apart, who'd never worked with me or Daktronics before.

"I'd start early each morning to be sure they got moving and understood the parameters and the timeline. In Texas high school football, expectations are high and everyone in town, from grade-schoolers to grandmas, attends the games. Failure was simply not an option.

"During that week, I guided both installation crews through the mechanical and electrical installation processes, helped pull signal cable, terminated fiber optic connectors (the old hand-polish type – very tedious work), and performed operator training on our Venus 4600 control software to groups from both schools. I put in some quality windshield time going back and forth to ensure that no details were overlooked."

But as with most installations, things happen and Wildeman found himself with some engineering issues that threatened the completion of the board by the rapidly approaching, Friday night game. The athletic di-

Xtra

Winter 2008 **DAKTRONICS AQUATICS NEWS**

BUILDING LEGENDARY SERVICE IN TEXAS

According to former swim coach Steve Montgomery, "Texas aquatics coaches are realizing Daktronics is here to stay." Last fall, Montgomery and his colleagues from the Texas DSS offices once again setup a booth at the Texas Interscholastic Swim Coaches Association Clinic in Austin. "For the second year, we saw lots of interest and inquires. Coaches are realizing that our service is available, and that is of great importance."

From Houston's Lamar High School to Austin's beautiful Texas Swimming Center on the UT campus, Daktronics timing and scoring system and service is becoming popular throughout Texas swimming.

Mike Waldmann of Andrews High School in West Texas is one of those satisfied Daktronics users. "We were truly excited about our new pool and equally excited about the new timing system. The instant feedback from our workouts is wonderful instead of having a coach with his stopwatch try to keep up with who came in and when.

"Andrews can now host efficient meets. No more writing swimmers' times on a board from hand-held stopwatches. More importantly, no more of those dreaded discussions after a close race." Waldmann added, "We made the decision to use Daktronics because of customer service. Being in coaching for the past 24 years, I have been exposed to other timing systems that worked well when they worked. As I listened to coaches share the issues they've encountered, it was clear that most concerns centered on service. With an office only two hours away in Lubbock, I felt that we would be well serviced. And we haven't been disappointed."

Swimming legends are still being made in Texas and Daktronics will be there.

The Texas Longhorns installed their new display. Watch for the story in next spring's THE LATEST SCORE.

Lubbock I.S.D. installed a new GalaxyPro® display in their Pete Ragus Aquatic Center along with the OmniSport® 2000 timing system.

Fort Worth's Texas Christian University pool was upgraded with a complete Daktronics scoring and timing system. Head Swimming and Diving Coach Richard Sybesma said, "We are ecstatic about our new Daktronics scoreboard and timing system at TCU. I am impressed at the ease of use and reliability of the timing system and relay judging platform. We have truly enhanced the fan experience at our swimming and diving meets with the graphics and team information that can be displayed." TCU's display shows both swimming and diving results, animation and short video clips in 68 billion colors.

Andrews High School numeric scoreboard.

rector, Pete Vela, came to Wildeman with concern and clear concern that the board would not be ready. Wildeman comments on the conversation with Vela,

"Honestly, I didn't know exactly what to say, but I remember saying something to the effect that, 'We're Daktronics. We deal with these high-stress situations all the time. It's the nature of our business. Your scoreboard will be running.' And most of me really believed what I was saying, even though *I'd* never been through this kind of situation before, with so much at stake. I guess I just believed in Daktronics and I was part of Daktronics."

And Wildeman, along with the Daktronics work ethic, and some engineering creativity, was able to pull off both of the installations by the game nights. "I had pulled it off by myself, but I never wanted to have to do that again," said Wildeman.

Wildeman slowly began to add staff on an as-needed basis. He hired students from local universities and technical institutes, four of which have become full-time Daktronics employees. Scott Houser, Chris Mangus, Jose Rodriguez and Maxine Lowery were four of the employees that Wildeman hired as students who have taken on full-time jobs.

"The San Antonio office has come a long way since its humble beginnings. It has gone from being a one man show in 1994 to a 14 person operation with other offices established in Houston, Dallas, Lubbock, and Weslaco."

Weslaco, Texas

In 2007, the Weslaco, Texas, office was opened by Osiris Garcia. The office now has four employees selling and servicing to the Weslaco area. Some of the projects that the office has been a part of are Weslaco Independent School District (ISD), Harlingden ISD, the Ford dealer in Harlingden and Sherryland ISD. However, there are a few projects that have remained very memorable for Garcia and his team.

One of the more embarrassing ones for Garcia was while working on a project in Weslaco. The customer got frustrated with Daktronics and took out some of their frustration on Garcia by "chewing him out." Being a new employee to Daktronics, and taking over towards the end of the project, Garcia was unsure how to address the situation. The only thing he knew to do was listen to the customer and respond to the issues at hand.

After all was said and done, the customer ended up changing their minds about Daktronics and Garcia. Garcia has worked hard for the customer by remaining available and returning phone calls to provide a high level of service. Whenever there are issues, he and the office resolves them. Since that initial install, the Weslaco office has been able to sell the same customer multiple other displays. "We have been able to deliver on the service we promised," said Garcia.

While working on an installation at Sherryland Independent School District, Garcia kept running into issues with the display. He and the electricians were working late into Thursday night and Friday morning to get the display ready for the Friday night home game. However, they could not get the sound system within the display to turn on and work. After several calls to the corporate office in Brookings, Garcia realized that he had

Osiris Garcia

Pete Vela

Weslaco Independent School District.

forgotten to connect a cable within the system. At the time, it was a highly stressful event for Garcia, but now he is able to laugh at the situation.

Garcia can attribute the success of the office to a few people. The Houston, Texas office has been a huge help to Garcia. They have acted as a mentor for the smaller office and has been there when Garcia needed some extra assistance. Another contributor to the success of the office is Pete Vela. Vela sells displays for Daktronics through the Retired Athletic Director program. His knowledge of the equipment and the customers has been foundational in helping the office succeed in the Weslaco area.

Great Lakes Region

Indiana

Service Pays Off in Avon

A long term goal for the Avon, Indiana, Daktronics Sales and Service office had been to break a strong allegiance between the Carmel Clay Schools district and a close competitor. John Paloma, office manager, says that Carmel Clay, one of the wealthiest school systems in the country, used to be the competitor's "poster child." But a lack of service changed all that.

"I remember when I was working with Darrell Thiner and asked if he ever contacted Carmel. He said he had tried, but with no success. I said 'What the heck,' and headed out there. I had called the athletic director and left a message saying that I was going to be in the area. I was there for about three minutes, signed in, and the next thing I knew there was someone escorting me out of the building."

John had gotten his foot in the door, for a brief moment, before he got his feet thrown out the door. But this rocky start may have been the reason John received a phone call from the school a couple of days later. "What had actually happened was that the athletic director's wife, from Carmel Clay Middle

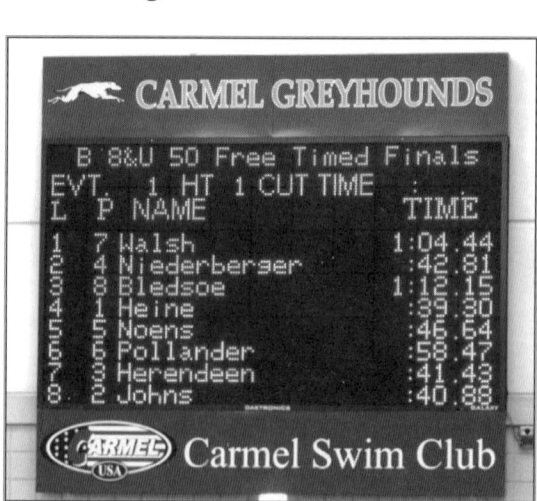

Without John's persistency and service, Daktronics may not have been able to break into the Carmel Clay School District.

School, had spent about seven thousand dollars on the competitor's service. She kept sending parts back, putting them in the board, and it wasn't working. That went on for a couple weeks before she called me and asked if I could come out and take a look at the scoreboards. 'I know we have the competitor's, but if you could come out and help with this scoreboard, I'd be so happy,' she said. I told her I was so busy right now, but I'd be there in an hour. I went over there to troubleshoot her scoreboard. She had two

Xtra

Winter 2005

DAKTRONICS AQUATICS NEWS

PALOMA TEACHES SERVICE FIRST

By Barbara Kleinjan

Customer service starts with respect. When you respect the customer and honor his or her right to be treated fairly and honestly, everything else is much easier. Daktronics Sales and Service Indianapolis Manager John Paloma begins his customer relationships by visiting the sites, talking with coaches and athletic directors, demonstrating the equipment, and sharing the opportunities that a Daktronics system can offer.

However, what really builds respect are his offers to provide training, at any time, to coaches and timing operators, to attend any swim meets necessary to trouble-shoot, teach preventative maintenance, and to respond to service calls anywhere in the state. According to Steve Maxwell, head swimming coach at New Palestine High School, "John went so far as to allow us to schedule a class for the parents from the high school, middle school, and swim club so that anyone who would be using the system was trained."

Besides promising fully operable installation and training, the DSS Indianapolis team tries to look for and teach preventative measures, but "when they need us, we come out no matter how simple the problem. We are always available," said Paloma. Customer satisfaction, "our greatest asset,"drives John's team of Randy Siebert and Mike Unger. The Indianapolis office team maintains that installation and operations need to be "not just done properly, but done well."

After finishing the wiring, DSS Indianapolis Manager John Paloma pots an in-deck plate at the pool in Brownsburg High School, Brownsburg, Ind.

Yet, emergencies do happen. When lightning hit the Mooresville High School aquatics facility, Coach Dennis Davis observed a true test of Daktronics resolve.

"You never really know what kind of service you get from manufacturers until something does go down — our local service is great," Davis added, "plus the Dak guys are really only a phone call away."

Reinforcing this, Maxwell added, "John is Daktronics to the people he comes into contact with. His professionalism and willingness to make it right carries a lot of weight when it comes to upgrading an existing system or purchasing a new one. It's like John understands that every old customer is just a word of mouth away from creating a new customer."

When New Palestine High School built its first swimming pool, Coach Maxwell cited two main reasons: product reliability and local service, for making Daktronics the only real choice. Maxwell remarked, "I never hesitate to recommend Daktronics; but I think I also feel I am recommending John Paloma, as well."

Daktronics Sales and Service Indianapolis

John Paloma

Randy Siebert

Mike Unger

331 32nd Avenue, PO Box 5128, Brookings, SD 57006
Phone: 800-325-8766 or 605-697-4300
Fax: 605-697-4700 www.daktronics.com
email: thelatestscore@daktronics.com
Galaxy®, OmniSport®, ProStar®, Venus®, are registered trademarks of Daktronics, Inc.
Copyright © 2005 Daktronics, Inc.   SL XXXXX 010405

Want to share an experience you have had with your local service provider? Email: thelatestscore@daktronics.com.

Go to www.daktronics.com/aquatics to complete a survey to win a HS-200 Horn Start.

output amplifiers on the competitor's drivers that were bad. 'That's what I needed to know,' she said. So she called the scoreboard company and said that they needed to repair some products. They told her that since I was from Daktronics I didn't know anything about their equipment. (It just so happened that I serviced many competitor's boards.) She called me up and said she had talked with the competitor and they didn't know what I was talking about. I wrote down 13 things that I checked and faxed it to her. She called up the competitors and told them what I did. 'Did he check this? Did he write this?' they asked her. They were going through all this and she started to ask them, 'Is this a good thing to check? Is this a good thing to check?' on all of the points. All the things were good things to check. She hung up on them and said she was done talking with the service and that she wanted to talk with the architects, because she was going to order Daktronics' equipment. I was very happy. I remember the day we got the purchase order. I sent an e-mail to Darrell Thiner and told him 'You'll never believe what I got." The service call really paid off...it was a great return on our investment."

Turning Point

Mooresville High School, about 30 minutes from Avon, Indiana, was refurbishing their entire school. This included a whole new swimming facility, complete with scoreboards. The bidding process wasn't entirely finished when John Paloma took a chance and taught the school what local service is all about.

"We were bidding against a competitor for the Mooresville High School project. They were building a new pool in their refurbished high school. Everything was set up for the competitor's boards. The school happened to be in-between coaches, so I started talking to the Athletic Director.

"What are you guys leaning towards," I asked.

"Your competitors," he said.

"You know what? I know you've been in communication with them, but how many times have you seen them?" I asked.

"I've never seen them," he said.

"How many times have I been here... talking about local service?' I paused for a second. 'When you get the competitor's equipment it's gonna look pretty, but something's gonna go wrong. If you were to get Daktronics' boards, number one...we always honor our warranty, because number two, we'll always be here. All the time."

"All the time?" he asked.

"All the time."

I think this was truly the turning point in our office for aquatics; it was what put us on the map. We used that high school as our marquee location. We took care of them and they let everybody else know how great of service our office provided. After that, we just started booming with aquatics."

"John Paloma (DSS Indianapolis manager) went so far as to allow us to schedule a class for the parents from the high school, middle school and swim club so that anyone who would be using the system was trained."

Steve Maxwell, head swimming coach at New Palestine High School (Ind.).

"It all started with you..."

"Indiana University has competitor's boards all over the school. I remember when I first got to Avon I was a service tech in the office. The assistant Athletic Director, also the voice of Indiana, called Darrell on a Thursday or Friday night before a football game and told us he was having problems with the scoreboards. He said that they might have been hit by lighting. Darrell told them that he would have to check and see what I was doing because it was last minute. He gave me the option to head on out there or leave it alone since it was a competitors. I was there all day and all night helping their

Indiana University

tech, it actually took a lot of hands-on, component level trouble shooting. We actually had to fix their drivers and their controllers because they had lightning damage. The Athletic Director really appreciated it. The way I know he appreciated it so much is because he bought all kinds of Daktronics equipment for Indiana University. In August 2006, we actually retrofitted their football scoreboard from incandescent to LED. One day I went out and was getting ready to do some work on one of their delay of game timers when I ran into Chuck. 'We sure have a lot of Daktronics here. You know, it all started with you.' I said it didn't actually start with me and he said that it was. I said it wasn't all me and that we have a team of people here to take care of you. He told me to stop being so humble. 'I'm just telling you that it all started with you.' Just goin' out there to help him, he recommended that everything go to Daktronics."

"Right Down the Road"

Plainsfield High School was building a new school, which needed new scoreboards to go along with it. John Paloma's persistence eventually won the contract.

"The Athletic Director from the current high school liked us. The Associate Athletic Director liked us. But the Superintendent at the facilities liked the competitor. We had the big guns cheering for us, but the guy with the mighty pen didn't like us. In the movie *Hoosiers*, the guy who made the winning shot with four seconds left in the game was Bobby Plum. He doesn't work for the competitor, but in his spare time he tries to get facilities to buy from them. He was trying to get Plainsfield to buy from our competitors for their new school. I became good friends with a couple of school board members and told them the truth; that our competitor is a good scoreboard company, but I said look at us and all the accolade we've got. We've been in the NCAA and the Olympics. The thing that really is going to top everything off is that we're right down the road from you. That's going to be the biggest difference because our prices are going to be the same. The big difference is going to be local service. They talked to all their school board members and the Athletic Director and ended up convincing the Superintendent to go with Daktronics."

Famous Encounters

One perk of working at high profile stadiums is the chance occurrence of running into a famous celebrity. This happened to Paloma when he ran into his boyhood hero in Indiana.

"I met Joe Montana at the Notre Dame-USC game back in 1997. We were in the perch on the top of the scoreboards. Right above the control location is very limited seating. A lot of family go up there and lounge around. He came in there with one of the priests. I said "Hi", and let him know we were part of the scoreboard service crew–my heart stopped. I've been a Notre Dame fan forever, since the 70s. It was so incredible, totally, totally awesome. Here's the Super Bowl MVP walking right next to me. I was like a little kid."

"I met Bobby Knight at the Indiana University Assembly Hall. I was just looking at the photos of the past teams. On all of their walls they have pictures of all the teams that played over there. I was looking at one of the 70s teams that went 32 and 0. I was just looking at it and minding my own business when Knight came and stood next to me. He talked about how much he loved coaching the undefeated team. 'That one was one of my favorite teams,' he said.

"Why was that?" I asked.

"Because they all listened to what I said. Everyone was gunning for us and they still came through."

"I couldn't believe this guy was talking to me in the first place. My heart was beating a thousand times a minute," commented Paloma.

Indiana University Basketball

Eating Crow

One of his most embarrassing moments happened to Paloma when he was servicing a scoreboard at Fulton Jr. High School; which was next to one of the biggest High Schools in the state. "They wanted me to take a look at a competitor's scoreboard for a game in a couple of days. As I opened up the back lid, a crow flew right at me. I just see these beady eyes coming at me; I was in complete terror mode and fell backwards – maybe ten feet to the ground. I remember brushing myself off, doing the old look around to see if anyone was watching. I continued working and acting as if nothing happened."

Kentucky

"...doing much better" in Lexington

In June of 2004, Scott Whittington moved from Indianapolis, Indiana, to the Horse Capital of the World, Lexington, Kentucky, to start Kentucky's first Daktronics Sales and Service office dedicated to serving Daktronics' customers.

"Initially, I worked from home to provide service, develop new and existing customers, and increase our market share within the state," said Whittington. "Many schools were using a lot of our competitors and had never heard of Daktronics, and our first year of sales proved that fact."

Not every sales call goes as expected, and being in a new area Scott Whittington had his share of rough experiences. "Everyone has had at least one bad cold call experience. I'm definitely no exception to the rule. My "all-time" bad cold call happened when I first moved to Kentucky in 2004, working from home as a "Man in the Van." I arrived at a county school in the morning, and checked in at the front office like I always did. I asked to speak with the athletic director and was taken down to the class where he was located. He had a study hall class, and seemed to be somewhat busy (usually a bad sign). In the back of my mind, I was thinking this isn't the kind of cold call I signed up to do.

Turfway Park, February 2008. Left to right: Scott Whittington, Amy Lawson, Stephen Edmondson, Ted Puthoff, Meghan Puthoff, Jason Vogel, Teresa Kinney and Steve Kinney.

I quickly introduced myself, the company I worked for, and why I was there. The introduction went like this, 'Hi my name is Scott Whittington, I'm with Daktronics, a scoreboard manufacturer, how are you today?'

'Well, I'll be doing much better when you're gone,' he replied.

Kentucky office

At that point I was pretty stunned because I couldn't believe that just happened. He didn't talk in a low voice, so everyone in the room heard the entire sales call (if you could call it that). That cold call was an introduction to the harsh reality that I had my work cut out for me in Kentucky."

'Win-Win' Situation for Retired Athletic Director

Whittington worked from his home office for about three years. On April 15, 2007, after structural changes and cosmetic upgrades, the first Daktronics Sales and Service Kentucky office opened. Steve Kinney was hired February 26, 2007 as a part-time salesperson to cover Louisville, Kentucky, along with the counties west of Interstate 65. Kinney is a retired athletic director from Ballard High School, located in Louisville, Kentucky, who has spent 34 years in the educational system. During that time he was a head basketball and baseball coach, assistant football coach, and Athletic Director.

"Steve has a great reputation and is highly regarded among his peers. He has proven to be a great asset for Daktronics," said Whittington.

"After thirty-four years in athletics as a coach and Athletic Director, I wasn't the 'new kid on the block' so to speak. Many of the coaches I had coached with had moved into Administration and other positions, some as Athletic Directors. When I became a sales-rep for Daktronics, it made the cold calls more like a visit than a sales presentation. The visits to schools became a way to stay in touch and talk 'athletics' while at the same time introducing Daktronics to new customers, a win – win situation."

The Importance of Install Numbers

It didn't take Steve Kinney long to learn that sometimes sales come from the unlikeliest of circumstances.

"The Athletic Director and I were friends and he asked me what I was doing and I said I just started working part-time for Daktronics. He mentioned he needed a scoreboard and we sold him what he wanted, a new scoreboard without install numbers."

(The importance of install numbers is a turn-key quote, which means that not only did the customer purchase a scoreboard, but Daktronics is responsible for the installation of the product; from the structure which supports the scoreboard to the electrical hook-up. Some customers feel the install numbers are too expensive and try to save money by installing using other means, which in this case proved costly instead of savings.)

"He retired that spring and a new A.D. came on board in July, so I dropped by to introduce myself and check on the new board. The A.D. was out on the hockey field helping the board of education employees dig the holes for the footers. When I arrived, water was gushing from the hole where they had just hit the water line to the sprinklers. I wanted to leave, but decided not to. The A.D. proceeded to tell me that this wasn't the first botched effort, as they had already hit the electric line leading to the football stadium. I wanted to say maybe you should have gotten the install numbers after all, but decided to keep my mouth shut."

"Where the old one was..."

Since everyone is human, everyone is prone to make mistakes. Steve Kinney talks about one contractor's mistake that resulted in a sale.

"A remarkable experience was in Spencer County High School with two nice young principals in a booming community, growing very quickly. They needed a quote on a new board for the back gym. We walked through the hallways into the freshly painted back gym. There was a new paint smell. I said where do you want the new board? He replied where the 'old one was'...which I looked around until I found a completely painted-over old board sitting on the floor. The contractor who painted, went right over the scoreboard that was on the wall."

Michigan

Leaving a First (and Lasting) Impression on Grand Rapids, Michigan

After graduating from South Dakota State University in May of 1995, Jeff and his wife, Heidi Gullickson, got an opportunity that not many people get; opening a Daktronics Sales and Service office in uncharted territory. Jeff opened the Niles, Michigan, office in August of 1995, and was in charge of everything – from sales and service to installations. It didn't take Jeff very long to make a lasting impression on Daktronics.

"I backed the van up to the overhang of the office. Probably because I was nervous going out on my own to a new area," said Jeff.

Jeff worked seven years as a one man show in Niles, Michigan. Each year, Jeff strived for more sales while delivering quality service. The first year Jeff sold $250,000 worth of Daktronics' equipment and increased by $100,000 each year until he reached his top sales of $1.1 million.

Jeff Gullickson and Carol Niles (2000).

Lasting Traditions

For the most part, Jeff serviced northern Indiana and the southwest corner of Michigan. This area includes Notre Dame, an independent college built on over 150 years of tradition. Part of the tradition was keeping incandescent, mechanical scoreboards. Jeff understood what they wanted and played a key role in getting the Notre Dame football stadium contract.

"The challenge with Notre Dame wasn't to present a big package. We wanted them to be a leader while they wanted to stay with very traditional gear."

Working on larger arenas is nothing new to Jeff as he has serviced the Pistons, Red Wings, Yost Arena, and the University of Michigan hockey arena. Getting Daktronics' name into the large arenas helped establish the company as a dominant force in other markets. "Pretty much when I moved into the area all the schools had older electrical mechanical equipment. The competitor had the stronghold in the area. Today, we're clearly the dominant scoreboard supplier."

Sales Made Easy Through Service

Fixing competitor's scoreboards was a regular part of Jeff's life and helped further establish Daktronics in the region. In fact, fixing a competitor's scoreboard helped Jeff get his first sale.

"Sales were easy when you serviced customers. Customers bought quality from Daktronics but they bought from the person who would take care of them," said Jeff. "I went in and fixed the boards. It was the first time in two years their electrical mechanical board was working. Two weeks later he ordered new boards from me. The Athletic Director became a good friend and I still get Christmas cards from him."

Continuing to Move Forward

Scott Houser replaced Jeff Gullickson in 2005 and moved the office from Niles to Grand Rapids, Michigan, and has run a million dollar office ever since.

"We are relatively proud that every school in the Battle Creek district has a ProStar® display, either in the pool, or out in the front of their auditorium."

Part of the success Scott Houser has obtained comes from moving a traditional customer into the future of LED boards. "Notre Dame is a high profile customer with a high level of expectation and we have a great relationship with them. The growth and level of service over the last four years has improved and proven that Daktronics is a trusted brand. We recently installed an LED baseball scoreboard, so they are moving out of the incandescent technology. There's a lot of tradition at Notre Dame, they have a lot of older alumni and they don't like to see change. Slowly but surely, they're changing."

Night Calls Provide Sales

Daktronics Sales and Service offices pride themselves on the level of professional service they provide. Whether it's being called the day before or weeks in advance; the offices get the job done. Scott Houser describes one circumstance that kept him working late into the night.

"Every year, during the football season, we have one or two. Every football season we're always working until late in the evening. Many customers don't appreciate it, but the ones who see it do. I remember two seasons ago, I was working in northern Indiana when I got called to a school in southern

Michigan. The Athletic Director was there and we got his scoreboard fixed. After that, we looked at the soccer field. It was just a field...no lights.... no nothing. So, we ended up driving my van and his truck and shining the headlights on it. I found out that a connector wasn't put in right on the back side of the display. It was about 12:20 a.m. before we left. When it was time for the school to buy new scoreboards, he [Athletic Director] didn't get any bids, he just submitted it. We've done a lot of good things by just providing standard customer service. Mainly, we provide excellent service on standard products. The same level of service we give to the larger installations, we give to the smaller as well. One of the things we like to say is 'Once they've seen the customer service we provide, we'll definitely get the sale.' The competitors are fairly close, but as far as service goes, hands down we can beat them on the service side of things."

Diane VanLeeuwen

Scott Houser

Notre Dame baseball scoreboard.

Past Life

Working as an office coordinator for about four years, Diane VanLeeuwen has received some interesting calls; but none that can quite top this experience.

"There was a guy who called in and insisted that we had weighted jump ropes. I've been here long enough to know what we have and what we don't have. The guy told me that he owned a local tennis facility and that he was such and such and it was now located on such and such street. He named the street which is strictly residential. He was very mad at me because I kept telling him that I was sorry, but we did not have jump ropes. He informed me that he had lunch with my manager, Scott Houser, and two of our sales guys at a nearby Chinese restaurant about two or three years ago and they discussed scoreboards and jump ropes. I got a hold of Scott and he said he never had lunch with the individual. This guy used to own a tennis facility in the area that went out of business. He's, apparently, at a retirement home 'living in the past.' He thought I was hording these weighted jump ropes from him. We've never had weighted jump ropes and we never will. He was apparently trying to set up a new facility at this retirement center. He took in different pieces from his past and put them together, fabricating info he didn't have or couldn't remember. He thought I was hiding them and not letting him have them. I would have been more than happy to give him them if we had them. To make this individual happy, I contacted two people at corporate that have been with Daktronics for 25 plus years to verify who we have never had jump ropes. I then called the gentleman back and he was perfectly content since my answer was 'verified'."

Consistency in Novi, Michigan

Novi, Michigan, has consistently reached their million dollar goal, year after year, in the HSPR market. Jason Snook, office manager, attributes the success of the office to the very large customer base in Michigan built from a relationship of trust in Daktronics' service.

"We're known for good sales and service out of this office. It's been very easy to work with a lot of people. Our reputation has been so good in the

Jason Snook

state of Michigan that our repeat sales are doing great. We've been able to turn customers into Daktronics customers from the reputation of our service," said Jason.

Another aspect of the Novi success included persistency while contacting customers. As Jason describes below, sometimes it takes years to get your foot in the door, but once you're in, the contracts can be very rewarding.

"The best project we've worked on is with Brandon Schools in Michigan. They purchased two football scoreboards, one is a standard board the other is a full-color message center...it took me two years to get the school. Now the Elementary and Middle School are looking for new signs. It was great to get into that new district and now we pretty much have everything. They're very happy with the scoreboards. Getting good compliments has been good all around. We're currently in the process of outfitting two districts around them."

Michigan School for the Deaf

Giving customers options for their scoreboards is a key part of the Novi Daktronics Sales and Service philosophy. Letting the customer know what's out there allows the office to pull in sales. Each customer has unique needs. Snook finds these needs and helps the customer find workable solutions. One such school with special needs was the Michigan School for the Deaf.

"The great thing about the sale was we sold them ColorSmarts® and Daktronics board light strips. Everything is very visual, lights are the type of buzzer. They were very happy with that. The fact that we were able to make their kids happy and not feeling alienated. They didn't think anyone was out there and we helped them," said Jason.

Sometimes It's Just not Your Fault

Grace under pressure is a valued quality among the Daktronics Sales and Service offices. There are times when technicians must understand that they're going to get yelled at for something they didn't do. Jason says that the ability to work with challenging customers is an everyday part of service.

"Service is huge since sales and service go hand in hand. The fact is that we come in and spend time with our customers. We serviced a hockey scoreboard. The arena managers were screaming and screaming, they needed it to be done as soon as possible. We crawled up there and found the problem. Prior to the game, a shot bounced against the wall, shot up into the air, hit the button on the power strip, and turned it off. All we had to do was turn the power strip back on. It was kinda' funny on our end. We were chewed out at first only to find that it was their problem."

Ohio

"Service Exceeding Expectations" in Columbus, Ohio

After serving as a technical operations manager for the PGA tour, Greg Denzinger opened the doors to the Columbus, Ohio, Daktronics Sales and Service office in the fall of 1994. "I came to know Daktronics from the customer side of the relationship. I was always impressed with the professionalism of their employees, even under pressure. With

Greg Denzinger

Broad and High Project

Columbus Crew Soccer Stadium

my engineering and project management background, I was equally impressed at how well they thought out the details of the projects. I also liked the company's view of the future. When it came time to see a new position, I knew I wanted to work for Daktronics."

Greg has worked on projects such as: Columbus Crew Soccer Stadium (first professional soccer stadium in the country), Nationwide Arena, and the Broad and High project. Denzinger has even had the opportunity to shake the hand of Jim Tressel (head football coach for the Ohio State University Buckeyes). "Service exceeding expectations" is Greg's philosophy for the Columbus office.

GREAT JOB!

Greg Denzinger, Scoreboard Sales and Service of Ohio, recently sold a 48 x 96 Galaxy® 3060 LED (light emitting diode) display to a Pepsi® bottler that he works with in the area. The display was installed at Wheelersburg Cinema in Wheelersburg, Ohio.

Denzinger estimates that 70-75 percent of their business comes from bottlers and feels it's important to establish and maintain relationships with them.

This sale is a good example of how Daktronics' regionalization process is working. Even though this was a commercial installation, the customer was able to work with Denzinger as the Daktronics' contact person.

"Regionalization has lessened the confusion and cleaned up the information and ordering process for our bottler. In this instance, they were able to do business with someone they already knew and were comfortable with. They didn't have to get to know another sales person and they know we'll take care of their customer just like we take care of them," said Denzinger.

A key to the sale was showing the customer an AF-3065 Galaxy® LED demonstration unit that Denzinger has in his office.

"It (the demonstration display) gave us one more opportunity to establish a personal relationship. The customer was able to more clearly define their needs in their own mind which helped us understand how to best help them," said Denzinger.

(From the January 31, 2002 "Extra! Extra!")

Copley, Ohio

The Copley, Ohio, office originally opened in 2000 as the Norton, Ohio, office working out of a small two room office location that was smaller than most normal sized living rooms in modern homes. One day Daryl Mihal, manager, looked around and noticed that they had more desk space than floor space. With the hiring of another technician the choice had to be made to either get rid of the bathroom or move to a larger office. The group, which started with three employees and grew to seven, moved five miles up the road to its current home in Copley.

"It was amazing to witness how fast the business exploded in northern Ohio. We had just won the Cleveland Indians Progressive Field project, Jacob's Field at that time, and were already supporting the Browns and many other high profile venues in the area. Along with the Large Sport Venue (LSV) projects, we had just sold six high schools video display systems within a few months and were already being told that we were going to be one of the test markets for the Clear Channel billboard displays. I

Cleveland Indians

don't think we stopped to even breathe for a few years," said Daryl Mihal.

Bob Dimichele, LSV technician, remembers the initial rush well when their planning varied from months to weeks to days to hours and even minutes before they realized what they were up against. "Things never really slowed down, they have all just gotten better at seeing through the fog of being so busy. Everybody has grown leadership skills that enable them to function as a true team that can handle the pressure of all the clients' expectations."

Daryl Mihal

The Copley office's success has not been fleeting, but has continued to grow. The staff reached ten in 2008 and will continue to grow as open positions are filled and the market continues to expand. The secret to their success is simple, honest, and straight forward; teamwork is the answer.

"From market based through regionalization to business unit focused and beyond all the ever changing three lettered acronyms we are famous for, we have all understood that our focus may

Bob Dimichele

change, who we each report to may vary, but we all answer to the team and each other first and at the end every day," says Kristopher Shrewsberry, field service technician. "Without teamwork we couldn't trust each other to do what we say we're going to do, when we say we're going to do it. None of this would work and we couldn't all finish what we need to get done on a daily basis."

The lead salesperson on many projects, Daryl Mihal, asserts that the sales process follows a similar dogma. He believes team sales have led to the success they have experienced.

Kristopher Shrewsberry

"From the coordinators and administrative help in Brookings who help keep our heads straight, to the project managers who keep our feet firmly on the ground and out of our mouths, and the technicians who help us deliver beyond our customers' expectations, every sale we do requires many touches to help it go smoothly from beginning to end." The sales philosophy is similarly simple in nature. "Give the customer what they want," attests Daryl. Instead of spending a majority of the time telling their customers what they need, they listen to the customers and then adding the "Wow" factor to exceed their expectations. With this method, customers are not able to stop talking about what a great company Daktronics is to their peers and friends. They also add that persistence is the key to the success they have experienced in selling large video systems to high schools and commercial clients in the area.

It is this persistence that helped them sell the largest video display to a high school in the country to Parma City School District, a public school in the suburbs of Cleveland. The sales process spanned over five years and included three different superintendents, two school boards, four athletic directors, and numerous roadblocks. When asked how they were able to sell Daktronics video and scoring displays to both sides of a historic football rivalry between Massillon and Canton High School, they said it was easy. Through their persistence and teamwork it is almost like clockwork. About every other year one of them tries to outdo the other. They sold the

The largest high school video display in the country at Parma City School District.

first video screen in Canton's Fawcett Stadium through a Daktronics Sports Marketing agreement. Massillon called a year later and wanted a video screen and they wanted it to be one module taller and one module wider than the one in Canton. The next year, Canton upped the ante and added another video screen to the other end of their stadium. What's next?

"Stay tuned...," says Daryl Mihal, "The commonality they share is their demand for excellence through every aspect of their football programs to their stadiums. You have to continue to deliver excellent product and services year after year because you always know the next project is just over the horizon."

Not only has persistence and drive led to success in large projects, but, it's the reason they point to for their many other successes in all standard projects.

Everyone in this office understands that the phone doesn't just ring. They make it ring. There is a lot of competition in the area for the large and smaller projects and they make sure to focus on driving the business to the clients through customer visits and relationship building. When the competition expects the 800 pound gorilla and they have to make sure to deliver it. At the same time they make sure to keep the pressure on their competitors and not the customers.

New England Region

Connecticut

Giant Repair

The Rocky Hill, Connecticut, Daktronics Sales and Service office opened late 2003 by Clark McAdams. Since then, the office hasn't moved, but it has expanded into the former-neighbors office. Working out of the "Constitution State" McAdams says that the Rocky Hill office is not your "traditional" Daktronics Sales and Service office, because their focus is mainly on service.

Rocky Hill may not be your typical Daktronics Sales and Service office but Clark McAdams regularly carries on the tradition of quality and dependable service. A testimony to that statement is his involvement in an emergency repair to the Giants Stadium display.

Giants Stadium in East Rutherford, New Jersey, was hosting the Major League Soccer season opener between the MetroStars and Real Salt Lake on April 2, 2005, when a major rainstorm crashed down on the facility. The violent storm, which produced ferocious 48 mile per hour winds and torrential rains, forced the teams to declare a scoreless draw.

Mario Musa, Meadowlands technical director, and his technicians immediately inspected the large screen video displays and found the storm had damaged some of the display components. Mario contacted Clark McAdams, Daktronics Keyframe representative in Rocky Hill, Connecticut, and informed him of the situation.

Aware of the stadium's busy schedule, Clark knew what needed to be done.

Koreen Bjorklund

Clark McAdams

He made arrangements with Daktronics Service Coordination Center to have the appropriate exchange parts sent directly to Giants Stadium. He also contacted Kurt Decker from Decker Electric, a Daktronics Authorized Service Company, and asked him to meet onsite to help with the installation as soon as the parts arrived.

"The replacement parts were shipped to Giants Stadium, and Decker and Clark worked great together and got the system up and running," said Mario. When the displays were turned back on, everyone was happy to see them operate flawlessly again.

Giants Stadium (2001).

Simple Solution

Some solutions at the Daktronics Sales and Service offices are not lasting, but they are simple. Koreen Bjorklund recalls one of those solutions while moving into the new office space in Rocky Hill.

"One of the electrical contractors that I worked with felt bad that I didn't have a desk, so he used miscellaneous office supplies to make me a desk. Cardboard box top, cardboard edge protection strips from the file cabinets, and binder clips. They even made sure I had the appropriate office supplies on my desk....including a notebook with the following OSHA notification. 'This desk is NOT OSHA rated and can hold no more that five pounds' Needless to say, I held my laptop in my lap while sitting in my lawn chair with my soda in the arm cup holder. Classic!"

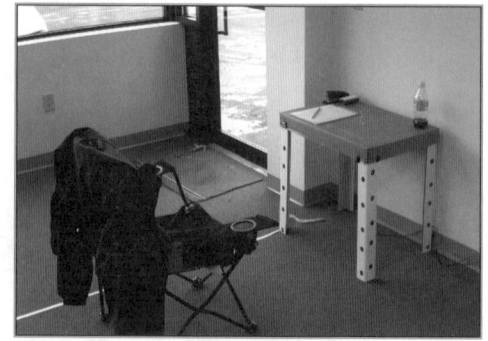
The legendary cardboard desk.

New York

"... a moment that I will never, ever, forget in Albany, New York."

Paul Farley opened the Albany, New York, Daktronics Sales and Service office in 2001. Farley has been involved in projects in major venues such as RFK Stadium (Washington D.C.) and FedEx Field (Landover, Maryland). But, throughout all of his years of work at Daktronics, there is one project that he will soon not forget.

"Last August, we had a parent who told us that his son died on the way home from a football practice and that he wanted to do something in honor of his son, Justin Wagner, a football player on the team. Justin was not just a good football player, but a gifted athlete. He played many positions on the football team and was being interviewed by a few colleges. On the way home from his practice in August of 2007, he got into a car accident and died instantly. His father wanted to put a video board up at the football stadium to honor his son and asked if there was anything we could do. In order to get a video board up by the last game of the season I contacted Greg Wilson, from the commercial market, and asked if I could borrow his trailer that had a Galaxy® Pro.

The final game was on a Friday at Mahopac High School. It was Wednesday, two days prior to the game, and we were having problems with the video board. We couldn't communicate correctly with the display by way of Ethernet. The Ethernet radios didn't work. This could have been in relation to the controller that was not working correctly. To make matters worse, it started to rain. This was not your normal rain storm. It came down hard, like a tropical storm. The school, understanding the pressure we were under, got us a tarp. The wind was so bad that the tarp was not good enough, and the rain started hitting our laptops and the inside of the display. When the rain let up (several hours later) we ran an external wire up to the press box. The only input that was currently working on the display was RS-232. It took over 35 minutes to download a small size file and we had 16 video clips to download. We estimated that it took over two days to download all of the content.

The second problem we had with the content was the size. All of the clips had to be downsized to fit our video board. We didn't have the software on-site to do this and if we made a request, it would take three to four days to complete. This was a problem since we only had two days to work with. Scott, Justin's Father, owns a network company and was able to get the software needed to downsize the files. Once we did a file, we realized it would take over a day to downsize all the video clips. Although the game was originally set for Friday, a rainy forecast postponed that game to Saturday. Midway through Friday, the Ethernet started to work, to this day we don't know why, and we were able to download most of the content. By Friday afternoon, we downloaded all the video clips of his son playing football, including an interview with a local TV station. This is also the day we had a spare part come in to fix our communication issues. We were able to download clips in just a matter of minutes.

The next issue we had on Friday was that the video board didn't have an output for sound. Scott Wagner wanted to play the sound during the video. We had a second laptop that had the sound software, but we needed to be in sync with the video. One of our major issues was that the sound clips were not the same lengths as the video. At this time it was 7 p.m. on Friday, but luck was on our side. Scott's daughter's boyfriend just happened to know the software that was being used and he chopped up the sound clips. He spent about an hour helping us get the sound clips to the same lengths as the video clips. Once he fixed that issue, we had one final issue; to get the sound to be in sync with the video. That required a lot of practice since we had to press play for the video and the sound at different times because it came from different computers. It took two hours to get it just right. I left the school at 10:00 p.m. and drove two and a half hours back home.

Saturday, the day of the big game, came. This was a very big game because it was their rivalry. Each year, the schools exchange a trophy named after a football player who died years ago. At the end of the game, Scott Wagner was introduced. Scott started to tell his son's story in front of 3,000 fans, parents, players, teachers, students, friends, and family. After Scott's speech we played the video; it was so quiet you could hear a pin drop. All the football players, cheerleaders, and coaches were sitting down on the field. Everyone in the stands turned to the end zone to watch this video. At the end, I don't think there was one person with a dry eye.

Everything went perfect. The video was in sync with the sound and everyone congratulated us for a great job.

Scott Wagner, his daughter's boyfriend, Scott's employee, Anthony Albano, John Ford and I put in almost 36 hours in three days to complete the presentation. I wouldn't have been able to do this without Brian Howk picking up my sales meetings.

Recently there was a district office meeting that started in the evening. This was to determine if the school will allow the video board to be installed. Unfortunately, they had a couple of things on the agenda and we were the 2nd part of the meeting. Upon leaving the district office meeting one of the board members told me not to worry. We got reassurance that the school was going to vote yes and Scott is going to get his video board. That meeting was set for March 11, 2008.

To this day, I really don't know how to explain the game being delayed which gave us the additional day we needed to get this project to work. Mother Nature? God? Did Justin Wagner have some type of presence? Or was it just plain luck? What are the odds? How often do you hear of a football game being pushed back another day because of rain? The Athletic Director had no idea of the issues we were having. He tells us about the delay on Thursday morning. Again, what are the odds?? What were the odds that the Ethernet port decided it was going to work on that Friday? The spare part didn't show up until Friday afternoon. If we had to wait, we might not have been done until way after midnight on the Friday. What were the odds for this to go right? Since we already had content downloaded on this electronic part, the controller, we made a decision not to change it. We didn't have the time to waste.

I found out later on that there were some complaints about the sound being played at 9:30 Friday night. The Athletic Director had told the few people who were calling what the situation was. We were never told to stop.

To Scotts Family,

I'm sorry for your lose. I can't begin to imagine the pain and suffering you have been through. That Saturday was a beautiful tribute for your son's legacy, for a sport that he truly loved and was gifted at.

I have to admit, that Saturday was a moment that I will never, ever, forget. I just wished it could have been for a happier occasion.

— Paul Farley

Paul Farley

New York City Office Thrives in Times Square

The New York City Daktronics Sales and Service office is located in the midst of Times Square, on the corner of 43rd and Broadway. David Ramirez, office manager, can look out the third story window of the Daktronics Sales and Service office and see the larger than life LED digital displays of Budweiser, TDK, and the NASDAQ Info Center - where Al Kurtenbach and board members stood under the NASDAQ ticker in 1994 for the

Dr. Al Kurtenbach, Co- founder and Chairman of the board opens trading on NASDAQ.

Times Square from the Daktronics Sales and Service office window.

NASDAQ ticker (right) from the New York Daktronics Sales and Service office.

first public offering of Daktronics stock. Currently, Daktronics has 15 high-profile digital displays in Times Square (approximately 1.5 million people visit daily). According to Ramirez, Daktronics' presence wasn't always as prevalent as it is today in the Square.

"Many years ago when we [Daktronics] first came into the square we had neon drivers (electronic controls that flash the messages on and off) and incandescent displays. When I came to Daktronics there were two displays that were LED. In the past eight years, all the other projects have come to be."

Gratifying experience, Spotlight Live Greatly Impacted from the Luxury of Local Sales and Service

Most of Daktronics' Daktronics Sales and Service offices are in the neighborhood of their customers. Ramirez points out that local presence allows Daktronics to go the extra mile for its customers. One such customer that benefited from Daktronics' proximity was Spotlight Live.

Spotlight Live

"Spotlight Live was really in a time and money crunch. We worked with them to suit their needs so one of our competitors wouldn't take it away. We pushed up the schedule and worked on the number to keep competitors away. We wanted to accommodate their grand opening, which we did."

Always Check Your Business Cards

David Ramirez has learned an important rule when meeting with high profile customers in a tense meeting situation; always check your business cards.

"We were getting close to a final decision on a project…As the meeting wrapped up we started handing out cards. I started handing out cards that said Jill Krogman. Needless-to-say, they were curious why a six foot man was called Jill. They were all looking at me and it was in a few moments that I realized why they were looking at me strangely."

David Ramirez

Coca Cola® Display

The innovative design of the Coke display in Times Square has become an international monument to America's modern society. Working down the street from the sign, David Ramirez saw the entire project from beginning to end.

"Watching how a huge corporation like Coke goes about this process was fascinating. For initial design, they had an open design contest for anybody who doesn't build regular signs. They wanted someone who was thinking outside the

The Samsung® display, located just above the Coca Cola® display on Tower two in Times Square.

The Coca Cola® display has become the symbol of the modern Times Square.

box. Coca Cola® whittled it down to four designs using focus groups that decided on the final decision – not some exec. The design was driven by that process. I've had a relationship with Coke for many years, so I had the inside track to get into the project. All of us will remember July Fourth when we turned on the Coke Display."

Painfully Long

Grand openings are often high profile events for companies boasting new Daktronics displays. Since the displays are on the front line of the company's image it's important that everything goes as planned; especially if a countdown is involved. David Ramirez will testify that not all events run as expected.

"We had worked on a countdown for their sign launching. When we started the countdown, it was supposed to be every second, from ten to zero. When we were waiting for the final event - the computer went into sleep mode. It was painfully long – doubled timing – it was really a bad moment since it took twice as long. "

As Ramirez will point out, the relationship between Daktronics and Samsung® was far from ruined from this minor mishap.

"The project went so smoothly that since then, Samsung® has purchased two more displays from Daktronics."

Mid-Atlantic Region

Maryland

Baltimore, Maryland

When Bryan Nagel heard Al Kurtenbach talking about the opportunities at a Scoreboard Sales and Service, he knew it was the perfect fit for him. So, in the August of 1992, Bryan loaded his family (wife and six week old son) into a U-Haul truck and headed to Seattle, with a final destination of Baltimore six months away.

"I spent the majority of my time in Seattle doing service work and getting used to the big city life," said Bryan. "It was a great experience working with my time in Seattle doing service work and getting used to the big city life." After his training in Seattle, Bryan moved to Baltimore in the winter of 1992 to open the fourth Scoreboard Sales and Service office. "Talk about a large task, trying to get to know an area in a few short days and determine where to place the office and your new home. With the help of Gary Gramm, we scouted multiple office sites and finally settled on one. Must have been okay, over the next 12 years I moved the office into two larger suites, but stayed in the same business park."

Paul Kurtenbach joined Bryan Nagel, Sandi Lutsche, and technician Mike Haggerty in 1998. Paul helps build trust with the customers and understands that 'service leads to sales.' The work Paul, Bryan, Sandi, and Mike did helped make Baltimore one of the most consistent Daktronics Sales and Service offices in the company.

"We were one of the first offices to

Sandi Lutsche

Paul Kurtenbach

Bryan Nagel (far left) and Mike Haggerty (far right) moving scoreboard.

reach a million... every year since 1999 (except one) we've been able to reach a million dollars," said Paul. "The local presence is the biggest reason... I've been out here for ten years now...there's not many customers that don't know who I am."

Service Scores at Ripken Stadium

Ripken Stadium in Aberdeen, Maryland, was founded in June 2002 and purchased ProStar®, Galaxy®, and BA-3724 for their fans enjoyment. Ripken Stadium has since become a popular minor-league ballpark thanks to baseball hero, Cal Ripken Jr. For some, Cal is no longer the only hero associated with the stadium.

"During the summer of 2004 they had been experiencing a few problems with the power supplies in the ProStar® and Daktronics had made multiple trips to repair the board for the games. I had spent numerous days at the facility, testing and repairing the ProStar® in order to get them ready for their long home stand in late July. I thought we had it fixed."

It was an early July morning when Paul Kurtenbach heard that it wasn't fixed. Kurtenbach was headed to Philadelphia "to see the sites" when he received a call from Ripken Stadium. He answered the phone and heard the panicked voice of the assistant general manager on the other end. He told Paul the signs were experiencing "technical difficulties."

"He said that the video board was down and the game was going to start in 90 minutes... he asked if there was anything I could do to help. My wife overheard the conversation and motioned for me to offer help. Luckily we were on I-95, just a few miles away from the stadium when he called."

When they arrived at the stadium, Paul's family was taken to the club seats where they were fed and entertained while he went straight to the ProStar® display. Kurtenbach worked diligently on a forty foot ladder as the crowd filtered into the stadium. He worked on the display during the opening announcements and introductions to find out the problems were being caused by a bad power supply.

"I managed to troubleshoot and repair the display just in time for the National Anthem. As I climbed down from the display I noticed that my family was nowhere to be found. I began to walk towards the control room to confirm that everything was functioning properly when I heard my son and daughter yelling to get my attention...from the box seats behind home plate. When I told them we could leave now, they didn't want to go. So we ended up staying to watch the Ironbirds play."

Daktronics thanks Paul for stepping up to the plate, hitting a home run, and sending the relationship between Daktronics and Ripken stadium into extra innings.

Presidential Service

In Chapter Five, "The White House Calling," Bryan Nagel talked about helping President Clinton make a speech about the national deficit using a Daktronics display. The other man involved in the event was Paul Kurtenbach; here's his side of the story.

"We had one job that Bryan Nagel and I were doing. It was an installation that started at 5:30 a.m. and was going to take all day to do...Corporate called and said they had an emergency. While we were doing an installation we had to go to the Dulles airport [Washington D.C.] to pick up a sign. The sign was x-rayed to make sure that it wasn't a bomb... we weren't sure if that'd be good for the sign. By 10:00 at night we were at D.C.... we're in the East Wing of the White House and we figured we needed to go back to our office and get some tools."

Bryan and Paul headed back to the Baltimore office at three in the morning and made some calls to the NOC. The display needed to countdown at a certain rate to display the drop in the National Deficit and show the Deficit change direction at a certain time during President Clinton's speech. Bryan and Paul made quite a few calls between three and four in the morning to make this happen. By 7:00 a.m. they were in the van and on their way back to the White House.

President Clinton's sign for his speech on the national deficit.

Paul Kurtenbach and Bryan Nagel after the successful speech and 36 hours straight work.

"We were up for 36 hours at this point. We had to take the numbers they gave us while we were being watched by the secret service and federal bankers. We were sitting feet from where the president was going to give his speech and sweating. We had to make the deficit change from negative to positive at the moment he was going to say a number...in order to do this, we had to calculate the numbers on the fly. We were able to make the right switch during the President's speech. After the speech we were sitting there talking and a guy grabs us and tells us that Bill Clinton wants to shake our hands...It was a great experience."

After a day and a half of work, Bryan and Paul were able to help the President complete his national address with a Daktronics display. They came through and delivered presidential service in a high pressure situation, for a speech that was broadcast across 60 channels nationally.

Pennsylvania

Philadelphia and Edgemont, Pennsylvania

In September 2005, the Edgemont, Pennsylvania, Daktronics Sales and Service office was opened for business. This came about because Lobec Inc, reseller for Daktronics in the Philadelphia area, ceased operations. Former staff members from Lobec were brought on to open the Edgemont office. Michelle DiRocco, HSPR Sales Rep, is one of those original members from Lobec. She has been selling Daktronics in the tri-state area for over ten years. The Edgemont office is located about ten miles from the Philadelphia city limits; conveniently located to service most areas in eastern Pennsylvania, New Jersey and Delaware. The office services the High School Park and Recreation (HSPR), Large Sport Venue (LSV), and Commercial Markets. Edgemont has two full time sales reps and one retired Athletic Director in north New Jersey, two full time technicians, two student interns, and one office coordinator. They are a small but powerful office that has achieved over $1 million in sales in the HSPR Market and $13 million in Commercial Market by their second year of business.

Michelle DiRocco

First in its Field

Most high schools across America are content with getting basic scoreboards, now schools are starting to invest in more technologically complex scoreboards. A milestone was reached in the HSPR market when Michelle

DiRocco sold the very first ProStar® display to a high school in the Mid-Atlantic region. "I have been selling Daktronics products for over eight years in the HSPR market and never thought I would have sold a ProStar®," said Michelle.

The Project: Hempfield High School in Landisville, Pennsylvania.

Hempfield High School

The Product: A 96 x 128 23mm ProStar® Video Display.

Galaxy® Proposal

Proposing can be a daunting task for any young man, especially when he wants to create a memory that lasts forever. Deb Conte recalls helping a student employee 'pop the question' using a Daktronics Galaxy® Pro.

"So one day Matt Nissley, our former student, who now works at Corp, calls me and wants to propose to his girlfriend Erica and needs some ideas. He was coming back to Pennsylvania for a wedding, his first thought was to be driving down the road and we would use our Galaxy® Pro to display "Will you marry me Erica," it was a great idea, but I didn't think we could do that with traffic laws. Then it hit me! Greg Wilson just installed a Gal-

The proposal

Matt and Erica

axy® Pro down in Ocean City, New Jersey, so Greg Wilson, Carol Sprenger and my husband Dave, with me planned a road trip. Greg set the sign up, Carol and I kept a look out for Matt and Erica.

As Matt and Erica were walking down the boardwalk (for those non New Jersey people, it is like a street but on the beach made of boards). As the couple approached the display, Matt stops, looks up and BAM! There it is, "MARRY ME ERICA", needless to say the bride-to-be could not stop crying...Matt and Erica currently reside in Brookings."

The Project: "Get Matt Hitched"

The Product: 80 x 144 Galaxy® Pro (owned by P.B. Weller, on the Boardwalk in Ocean City, New Jersey)

Did We Mention Edgemont has a Celebrity on Staff?

In December 2007, Tom Giorgio, our very own retired Athletic Director was featured on the front page of *Extra! Extra!* For receiving The New Jersey State Interscholastic Athletic Association's (NJSIAA) 2007 New Jersey Girls Tennis Official of the Year award. As mentioned in the article, Tom is also recognized in the New Jersey Coaches Hall of Fame as well as the Secaucus High School Hall of Fame.

Do You Swear to Tell the Whole Truth?

One of Daktronics' core values is helping customers, even if they're asked to go before a board. When Greg Wilson agreed to accompany a customer to court, for moral support, he got more than he asked for.

"A couple weeks ago I presented our products to the Office of Boards and Commissions for Newark, New Jersey. Originally the plan was to be moral support to my customer (who was there for moral support to his customer). But the end user's lawyer started the meeting out by calling me, the representative of the manufacture, to the stand to be sworn in. As I stood in the council chamber be-

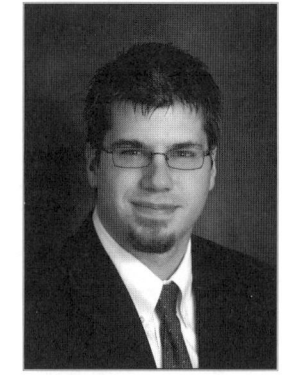

Greg Wilson

fore the board with my right hand raised the President of the board asked me if 'I solemnly swear to tell the truth, the whole truth, and nothing but the truth'. All I could think was 'Dear God, please don't let them ask me about shipping dates!'"

Extra! Extra!

Daktronics Daily Newsletter
Extra_Extra@daktronics.com

Giorgio Wins Official of the Year Award

The New Jersey State Interscholastic Athletic Association (NJSIAA) selected Tom Giorgio (HSPR) as the 2007 New Jersey Girls Tennis Official of the Year.

Tom Giorgio

"It's always an honor to be recognized by your peers and among the coaches," said Giorgio.

Besides officiating tennis, he also officiates volleyball and softball in New Jersey. Giorgio estimates that he referees around 75 games a year, and has no intention of slowing down.

"I hope to continue officiating, I enjoy it. It helps keep you young, and in touch with the youth and sports, at least mentally," said Giorgio. "And, as anyone who has served in both capacities will tell you, it is far less stressful than coaching."

Giorgio worked in education for 39 years before joining the Daktronics sales team in 2005. He currently covers sales for New Jersey and northeast Pennsylvania from Daktronics' Philadelphia office.

Giorgio will be presented with the award at the New Jersey Scholastic Coaches Association Hall of Fame Luncheon in April.

Previous honors include induction into the New Jersey Coaches Hall of Fame and the Secaucus High School Hall of Fame.

Daktronics congratulates Giorgio for his remarkable achievements.

Southeast Region

Florida

Clearwater, Florida, Daktronics Sales and Service

In 2004, Daktronics assigned an HSPR sales and service technician to the Keyframe office in Clearwater, Florida, officially transforming it into a Daktronics Sales and Service office.

Dan Roy, a technician for Live Events Services, has worked for Daktronics for seven years, and recalls several projects of which he is particularly proud.

Super Bowl XXXV in Raymond James Stadium – home of the Tampa Bay Buccaneers

"We have had some major dealings with the University of Georgia and University of Florida," said Roy. "We've also provided control room support for Raymond James Stadium where the Tampa Bay Buccaneers play. The most exciting event I have worked at was the Super Bowl in 2001. The Ravens and Giants played at Tampa Bay. It was really awesome to be part of that, and we plan to cover another Super Bowl there in 2009."

The University of Georgia chose Daktronics for several displays on their campus including ProStar® and ProSound systems for football, basketball, soccer, baseball, softball and volleyball.

"I like to think that every one of our customers is positively impacted by our sales and service," said Roy. "The University of Florida bought a vid-

University of Florida

eo board from us in 1998, but they weren't satisfied with us at first. But then Kyle Adams committed two technicians to service the boards for two years. The University ended up buying additional equipment for football and basketball. We worked hard, and continue to work hard to gain our customers' trust. Our philosophy is to go on site to show expertise and dedication to our equipment and dedication to customer satisfaction."

Dan Roy

Miami, Florida, Daktronics Sales and Service

In 2002, Bryan Clark started an Authorized Service Company (ASC) in his own home in Miami, Florida. Two years later, he opened the Miami Daktronics Sales and Service office. Today, Daktronics' southernmost Daktronics Sales and Service office has nine employees.

The office has sold and serviced Daktronics products for several large venues such as the Orange Bowl at the University of Miami, Dolphins Stadium, and American Airlines Arena, home of the Miami Heat.

After the 2006 football season, and early during the baseball season, a small problem was looming in the horizon for the Miami Daktronics Sales and Service office. At the same time Dolphins Stadium was scheduled to start a major renovation phase of expanding their "club" level.

"We started noticing some color differences between some of the modules in the 360 ProAd in Dolphins Stadium. Their color, for some reason, was not matching to the rest. Besides, five to ten modules was a small percentage out of the 5,300 modules that make part of the whole structure. At the same time we had to pay attention to the rest of the equipment. We thought that it had to be a small mistake in calibration from the manufacture department," said Victor Sarmiento, a technician for the Miami office. "As time went on, suddenly, the little problem was growing out of control. Now the Dolphin Stadium management was curious to what was going on, thus encouraging us to take action. As we began to work on the problem, we changed some of the modules but the problem was still there. After that attempt did not gave us the substantial results, we looked deep inside the boards itself. Data distributors, main line controllers, fiber, CAT5 cable, and power supplies were all methodically inspected and replaced, but nothing, not a single sign of success."

The technicians then contacted the Daktronics engineering department which created the ProAd display, but they still could not find a solution to the problem, which had eventually spread entirely throughout the north and south side of the ProAd boards. As frustration was building up, and no one could come out with a logical solution, on one sunny and sweaty Miami afternoon in May, a Daktronics student employee solved it as he was changing some modules.

"As he was coming down from the ladder be began to sneeze uncontrollably," said Sarmiento. "He decided to wipe off the modules because he didn't want to get sick. After the cleaning was done, and as he put the module back – eureka! - A sudden revelation was in his sight, and as a detective would, he started to put the pieces together. There was major construction taking place in both the north and south sides of Dolphins Stadium, which was creating incredible amounts of dust,

Dolphin Stadium

which was, in fact, settling in the lips of the modules and altering the color configuration. So after almost three months of chasing a ghost problem inside the equipment, it took a small sneeze to solve our problem."

Orlando, Florida, Office Helps Hurricane Recovery

In May of 1998, Steve Duncheskie opened Daktronics' first Floridian "Scoreboard Sales and Service Office" in Lakeland, Florida. After one year of business, Brad Flath moved from Brookings, South Dakota, to Lakeland to manage the office. The office found a new home in Winter Park, a suburb of Orlando, in the spring of 2000. In 2005, the office moved again to its current location in Orlando.

Mike Kempany has managed the Orlando office since 2004.

In 2004, central Florida was hit with an unusually harsh hurricane season. In the year after, the Orlando office's sales went from about $400,000 to $1 million annually.

This basketball display at Mainland High School in Daytona, Florida, was sold and installed by the Orlando Daktronics Sales and Sales office.

"It's never a good thing when the hurricanes come through," said Kempany, who now supervises nine employees in Orlando. "We were very glad to help out the schools and businesses in our area in the aftermath. Both Daktronics and our communities benefitted in the recovery efforts that year."

Kempany is particularly proud of the business relationship between Daktronics and the Disney Wide World of Sports® Complex.

"We've gone back and forth with them for several years now. They started out with our competitor's products, but then they realized that we have much better service to go along with our products," said Kempany. "Now there are about 33 Daktronics signs of all types in the Complex. We

Woodmont High School

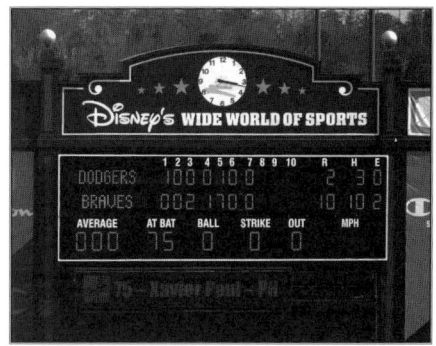
One of the displays at Disney's Wide World of Sports® Complex.

Vince Carter's curved DakTicker® in his lounge area.

also have Galaxy® products in Epcot, Blizzard Beach and other locations in Disney World®."

When Vince Carter, a guard for the New Jersey Nets, moved into his new home in an Orlando suburb, he ordered a custom curved DakTicker® for his lounge area and a scoreboard for his private gymnasium. The Orlando office was on the spot to provide

Vince Carter's scoreboard for his private gym.

the NBA star with quality products and service.

As the Daktronics employees were installing the DakTicker® in the Carter residence, they needed a computer to put the Venus control program on so that Vince could control the sign.

"Vince had this ancient Sony® laptop which didn't have nearly enough memory. I was pretty surprised," said Mike. "Me and another technician argued over who should go and tell Vince – he was sitting outside by the pool. Then I went out there and said something like 'this computer just isn't going to cut it.' Vince was pretty cool about it, and he said that he was getting a sponsorship from Macintosh, so he could take care of the problem, but our Venus programs aren't compatible with Macintosh computers, so Vince went out and bought a top-of-the-line Gateway computer just for the program to control his boards."

When the project was completed, he invited a few of the Orlando Daktronics Sales and Service employees to his home for a pool party.

"He was really cool and easy to work with," said Mike. "We chatted with him and his wife, and they were really social with all of the guests. It was a good time."

The center-hung display at the Master's Academy in Oviedo, Florida, was sold and installed by the Orlando Daktronics Sales and Service office.

Georgia

Alpharetta – An Agent of Expansion

In its third location, and with sales that exceed $1 million per year, the Alpharetta, Georgia office isn't a stranger to growth. The office started in 1994 with Bruce Chapman at the helm. He stayed for four years, and then moved to the Nashville, Tennessee, Daktronics Sales and Service office. That's where Rob Harrell came in.

A Matside II® sits in the Alpharetta office.   Alpharetta storage space in 1996.

Rob Harrell   University of Georgia

"When I got to Alpharetta, there was a manager, an office coordinator, and me, a technician," said Rob, who now oversees the more than ten employees in Alpharetta. "It really has grown a lot since then."

According to Rob, the biggest sales goal that the Georgia office has achieved was hitting the elusive $1 million mark in 2001.

"We quickly found out, after hitting $1 million, that it is hard to expand much more than that," said Rob. "Right around $1 million a year is where we top out with the amount of work that our employees can do."

When Daktronics introduced the Sportsound® product line, the Alpharetta office came out of the gates and sold more than any Daktronics Sales and Service office in the first year. They have also sold several notable ProStar® displays.

"As of 2007, we have sold about 15 ProStar's® out of this office," said Rob. "We also do event support for several ProStar® owners such as the University of Georgia (even before Keyframe), Georgia Tech, University of North Carolina, University of Tennessee, and the Hickory Crawdads minor league baseball team.

Phillips Arena

Rob's very first project in 1998 was for Phillips Arena, home of the Atlanta Thrashers NHL team, the Atlanta Hawks NBA team, and the Atlanta Dream WNBA team.

"I was particularly proud of that project because I worked 12-hour days for two months straight to meet the deadline," said Rob. "It was a very large and extensive project. I was there so long that I got to meet almost all of the Atlanta Hawks, and even some of the visiting players like Kobe and Shaq."

The Alpharetta office, although one of the farthest Daktronics Sales and Service offices from the corporate Daktronics office geographically, holds a philosophy that ties it close to home.

"You have to lean back on the philosophy that was given to us years ago by Frank Kurtenbach," said Rob. "He taught us that every sales opportunity is also a service opportunity. It is an opportunity to sell them something else – bigger, better, and something they don't have."

Close Calls

Despite the fact that the Alpharetta office has experienced great success in recent years, work in the office has not always been easy.

"The most difficult sale I had to do was for Dekalb County, the second largest county in Georgia. Daktronics lost a contract with Coca Cola® to sponsor scoreboards in Dekalb County. What made it difficult is that I had started talking with them five years earlier about Daktronics products," he explained. "Well I still held on, and the good news is that we eventually sold them 25 scoreboards. It paid off in the end, and it just goes to show that Daktronics is something that will continue to happen in the future if we stay dedicated to our sales and service."

The office has also come very close to deadlines on installations on a few instances.

"Starrsmill High School in Fayetteville, Georgia, was a close call," said Rob. "We had closed a deal on the display, and it was delivered so late that the actual board didn't arrive until the Wednesday before their football team's home opener on a Friday. I was working very hard to get that thing up and going. We got it up on the structure on Thursday, and the wiring was done at 2 p.m. on Friday. The school's principal and the president of the booster club were watching me scramble and taking bets on whether or not I was going to get it done. I had it done at 6:15 and the game started at 7:00 as scheduled. I had never felt so relieved in my life."

NORTH CAROLINA

Concord Services Daktronics 'Lifeline'

In July of 2001, Matt Lundberg moved to Concord, North Carolina, to set up a Daktronics Sales and Service Office. Since then, the office has been selling, servicing and installing several projects from small portables to 48-foot billboards. The office started as an HSPR sales and service office, and now, with a staff of 14, it covers all of the Daktronics markets.

Matt Lundberg

"The opportunity to work between every market at Daktronics has really opened my eyes to all different kinds of customers," said Lundberg, who heads up the Concord office and also serves as manager of the Daktronics Sales and Service office in Lexington, South Carolina, and HSPR Regional Manager for the Southeast Region. He attributes the success of the Concord office to their commitment to their philosophy on sales and service. "Our customers are the lifeline of the company and we're there to meet their wants or needs when it comes to our products. If it wasn't for the customer, we wouldn't even need our Daktronics Sales and Service offices."

The most notable project for the office in Concord thus far was for the Charlotte Bobcats Arena in Charlotte, North Carolina, and was installed in the fall of 2005.

Charlotte Bobcats Arena

"It was such a massive project and it took up a lot of time," said Lundberg. "It has been a showcase install for the region."

Another notable project for the Concord office was a digital billboard network for Adams Outdoor Advertising consisting of 18 Daktronics digital billboards.

The Concord Daktronics Sales and Service office also services Daktronics products for the Daktronics Sales and Service office in Lexington, South Carolina.

"We have a really good working relationship with the Lexington office," said Sharyn Turner, Office Coordinator for Concord. "It's a great situation; we cover for each other if we need help."

One of the eighteen digital billboards for Adams Outdoor Advertising.

South Carolina

Lexington Daktronics Sales and Service

The Daktronics Sales and Service office in Lexington, South Carolina, doesn't have as many sales or make as much money as many other Daktronics Sales and Service offices, but it does play a prominent role in the Lexington community.

"Our office was struggling when I started working here, but things got better," said Melba Shull, a 12-year employee. "It was hard at first, but we started making some progress and got some new customers."

After Scott Dieck, current manager of the Colorado Daktronics Sales and Service office, came in 1997, sales improved dramatically. The office made a considerable amount of income by servicing the competition's displays, and then began selling larger projects to area high schools.

"We recently sold a ProStar® to Byrnes High School for almost $350,000," said Shull. "That doesn't seem that big compared to other projects, but its huge for us in Lexington. In 1997, we weren't even making a profit, but within five years we hit $600,000 in annual sales. Now we're shooting for $1 million. I'm very proud of our office and how far we've come."

Melba Shull

She also noted the importance of the Lexington office's service.

"Our biggest asset is the local service," she said. "Our customers know that they can call us and get superior service locally. One thing that really helped us out in the beginning was our service of competitors' boards, and when the custom-

Byrnes High School

ers couldn't get the competitors' service, they turned to us, and eventually most of them got our products too."

Shull acknowledges Coach Evans, a retired athletic director working for Daktronics, for the progress in Lexington's sales.

"Coach served as the athletic director for Dorman High School, and he was a very nice guy to work with," said Shull. "He retired and now works for Daktronics as a salesman. He has been instrumental in getting projects that we would not have sold otherwise."

One of the biggest changes that Shull has seen since her start in 1996 is the development of the Daktronics product lineup.

"I still have my cheat-sheet from the incandescent displays and glow cubes," she said. "It's just amazing how much we have come up with since I started."

The Lexington Daktronics Sales and Service office continues to market the Daktronics brand in the community, and hopes to sell its first Sportsound® system in 2008.

"We're getting excited to sell some sound systems," said Shull. "Our customers should be excited about our products too. We have a tough market in this area, but the schools in our area are realizing that they can get more expensive boards by selling advertising to put on the boards. They get excited about the things they can do with Daktronics products. It just adds so much to a ball game when the boards are running cool content and working right."

One of the Lexington office's most prominent customers is a Pepsi® bottling company in Florence, South Carolina.

"They will only work with me, nobody else," said Shull. "I was talking to him the other day, he was ordering Pepsi® advertisements for basketball scoreboards, and I asked him which 'Coke' sign they wanted on the advertisement. He was like 'Excuse me?' because Coke is a swearword to those guys. I'm glad he has a good personality and is still a valuable customer."

Tennessee

Nashville Daktronics Sales and Service

In 1993, Bruce Chapman opened the Atlanta, Georgia office. Nine years later, Chapman relocated to Nashville, Tennessee, to open the Nashville Daktronics Sales and Service office. Today the office has moved to a larger rental unit and currently employs seven people.

The success of the office can be directly attributed to the office's selling philosophy. It is kind of like the song, make new friends but keep the old. The office strives to develop new customer relationships but always maintains current customer relationships. Because of this philosophy, the office sales have been increasing.

Bruce Chapman in Atlanta, Georgia.

Daktronics Sales and Service Installation

Daktronics Sales and Service Installation

Daktronics Sales and Service Installation

The Racquet Club of Memphis.

One of the office's first major equipment upgrades was to The Racquet Club of Memphis. However, this is the project that Chapman considers to be one of his most challenging. There were some issues going on with the board. He started working at 7:00 Thursday morning and continued working until Friday evening at 7:30. Hartman thought he had lost a valued customer because of the technical issues. However, his all-nighter proved to be successful for the following year as the club contracted for another display.

Epilogue

As we established our first Sales and Service offices, beginning in 1988, we were pioneering Daktronics in the respective geographical area with each new office.

The seeds of our sales and service presence first sown in 1988 have sprouted into a nationwide and international sales and service force outside of Brookings, numbering more than 500 full time, part time, and student employees, more than the total size of the company when we launched our first office.

We have come a long way. In 1988 Daktronics annual revenues were approximately 20 million, and our full time employee count was less than 300. In 2008, our annual revenues are 500 million, with full time employee count of 2,500.

Looking forward, the strategy of having many small bricks –and- mortar offices may change. Sales people are able to work effectively from their home using the internet. We expect that soon our service people will have wireless internet connections right in their service vans. The price of fuel will force us to be ever more efficient in how we deliver service. These are some of the opportunities and challenges that lie before us. But the importance of being able to serve our customers will not diminish.

As we look forward to what the future may bring, it is appropriate to reflect back on how far we have come, and to express our sincere appreciation for the hard work and commitment made by those who provided the leadership to open the offices that have grown into the sales and service infrastructure that supports our efforts around the country, and today around the world.

— Jim Morgan, President and Chief Executive Officer

Going forward we recognize the Daktronics Sales and Service office as an excellent way to attract students to the Daktronics team. Just as when Al and Duane started Daktronics to employ the talent supplied by SDSU, now our Daktronics Sales and Service offices are doing the same only regionally. Business and Technical students with new ideas and talents continue to fuel the growth of the company.

Serving our customers in today's world is challenging. Systems continue to get more complex and schools are using more volunteers to keep control of budgets. Our customers truly appreciate our talented employees located in close proximity to their schools and places of business. The internet will never replace having a local person stop in to lend a helping hand.

— Dan Bierschbach, V.P. Schools and Theatres and After Sales Service

About the Authors

Chuck Cecil was born in Wessington Springs, SD, and graduated from Rapid City High School. An aerial photographer in the Navy during the Korean War, he earned bachelor's and master's degrees in journalism from South Dakota State University. Cecil is a former news editor for the *Watertown Public Opinion* and editor of the *Vermillion Plain Talk*. He joined SDSU in 1965  as development and public relations director and later served as assistant to three SDSU presidents. He took early retirement in 1985 and built a weekly newspaper chain that eventually included the Estelline *Journal*, Volga *Tribune*, Toronto *Herald*, White *Leader*, Elkton *Record*, Dell Rapids *Tribune*, Baltic *Beacon*, Brandon Valley *Challenger*, *The RFD News*, and the Moody County *Enterprise* in Flandreau. He retired from the newspaper business in 2000 and now writes a weekly column for the Brookings *Register* and does other freelance writing. He has written fifteen books including *Pony Hills*; *Stubble Mulch*; *The RFD News*; *Becoming Someplace Special*; *A Brookings Album*; *Remember The Time*; *Myron Lee and the Caddies*; *Family Matters-You Can Bank On It*; *Going The Extra Mile: The Story of the South Dakota Rural Electric Association*; *Nick's Hamburger Shop*; *The Man Who Lit It Up*; and *Fire the Anvils, Beat the Drums*.

Carl Deardoff graduated from South Dakota State University in Brookings, South Dakota, in May of 2008 with a degree in broadcast journalism. He worked in the Employee Communications group at Daktronics for eight months before accepting an internship in the Daktronics Schools and Theatres business unit. Carl is originally from Baltic, South Dakota. While attending SDSU, Carl was a station manager for the campus FM radio station, KSDJ.

Sarah Even is currently pursuing a journalism and mass communications degree from South Dakota State University in Brookings, South Dakota, and will graduate in May 2010. She has held two summer internships for the state of South Dakota, including Public Information Intern for the Department of Public Safety and Communications Intern for the Governor's Office.

Lyndi Hawke is pursuing a bachelor's degree in art education at South Dakota State University in Brookings, South Dakota, and will graduate in May of 2009. Lyndi hails from Sioux Falls, South Dakota. She works as a graphic artist in the Employee Communications group at Daktronics. Lyndi's artwork is currently featured in the Ritz Gallery at SDSU. She is looking forward to being an art educator in the state of South Dakota post graduation.

John Nelson graduated from South Dakota State University in Brookings, South Dakota, in May of 2008 with a degree in news-editorial journalism. He served one year for Daktronics as a student in Employee Communications where he completed an academic internship. John worked three semesters as a photo editor and chief photographer for SDSU's student-run newspaper, *The Collegian*. After graduation, he accepted a full-time position as Daktronics' first employee dedicated to marketing customer service for all business units.

Jen Oolman graduated from South Dakota State University in Brookings, South Dakota, in May of 2008 with a degree in journalism and Spanish. In 2007, she completed her internship at Daktronics as a graphic artist for Recruiting and Training. After being employed here for two years as a student, Jen is now exploring her full-time career opportunities at Daktronics.

Caitlin Osborne is pursuing her bachelor's degree in English education with a minor in mass communication from South Dakota State University in Brookings, South Dakota. She will complete her student teaching in Watertown, South Dakota, in the fall and graduate in December of 2008. After graduation, Caitlin plans to teach at the middle school level and eventually pursue a master's degree as a school library media specialist.

Staci Perry graduated from Southeast Technical Institute in Sioux Falls, South Dakota, with A.A.S. degrees in marketing and in business administration. The University of Sioux Falls is where she earned her B.A. in business management. As she continues her education, Staci is currently attending South Dakota State University in Brookings, South Dakota, to earn her master's degree in communication studies and journalism. Staci worked as the marketing manager for the annual Bash in the Grass event in Volga, South Dakota, and at Trotline Marketing before opening her own marketing business, Jo Michael Marketing. From 2003-2004, she worked at Daktronics as a Sales Coordinator in the Commercial Market. Staci continues freelance work for seasonal businesses, the motorsports industry and nonprofit organizations. In 2007, Staci joined the Daktronics family once again and is excited to serve as the Communications Manager.

Index

1988 Inventory List, 29
1990 Goodwill Games, 36
Adams Central High School, 167
Adams Outdoor Advertising, 242
Adams, Kyle, 234
Airline High School, 176
Albano, Anthony, 219
Albany, NY, 77, 217
Albuquerque Public School, 127-128
Albuquerque, NM, 126-128
Alcorn, Ted, 140
Alrus High School, 87, 162-163
Alton School District #11, 111, 166
Alton, IL, 111
American Airlines Arena, 234
Anaheim, CA, 138
Andraschko, Scott, 106
Andrews High School, 190
Angels Stadium, 138-139
Ankeny, IA, 67-68, 71, 148, 151
Apharetta, GA, 238-240
Appalachian State University, 104
Arizona Diamondbacks, 7
Armory Foundation, 102
Arney, Mitch, 151
ASA Hall of Fame Stadium, 87
Atlanta Braves, 106
Auburn University, 96
Avon, IN, 194, 196-197
Avon, MA, 101

Ball State University, 68
Baltimore, MD, 61, 74, 159
Barquisimeo, Venezuela, 110
Bati, Dan, 140

Baton Rouge, LA, 77, 176-177
Battle Creek Public School District, 205
Becker Furniture World, 157
Bellagio, 126
Belleville West High School, 166
Bellevue High School, 40
Bertram, Joel, 158
Bethesda, MD, 78
Bettner, G. Robert, 101
Bierman, Jeremy, 170
Bierschbach, Dan, 50, 248
Bierscheid, Bob, 100
Billings, MO, 61
Bjorklund, Koreen, 215-216
Blankenship, Scott, 147-148
Bloomington, IL, 145
Blue Man Group, 125-126
Bobinski, Michael A., 98
Bonneville High School, 131
Boone, NC, 104
Bossier City, LA, 176
Bow, WA, 103
Brandon School District, 208
Branton, Tim, 61
Breyfogle, Steve, 105
Brighton High School, 116
Broad and High Project, 209-210
Brookings, SD, 2, 104, 123, 134, 172, 192
Budweiser, 220
Buendia, Carl, 139-140
Burke High School, 168
Busch Stadium, 160, 162
Butte, MT, 120
Byrnes High School, 243

Calgary Olympics, 31
Camden Yards, 76
Camp, Meredith, 100
Canton High School Fawcett
   Stadium, 214
Carbondale, IL, 144
Carlos, Juan, 139
Carmel Clay School, 194
Carmi, IL, 111
Carmi-White County High
   School, 111, 166
Carnahan, Scott, 96
Carr, Mark, 177-178
Carter, Vince, 237-238
Cecil, Chuck, 249
Central Catholic High School, 145
Chaifetz Arena, 161-162
Chaparral High School
   Aquatics, 141-142
Chapman, Bruce, 238, 245
Charleston, IL, 107-108
Charolotte Bobcats Arena, 242
Chicago Midway Airport, 146-147
Chicago O'Hare International
   Airport, 146-148
Cillo, Anthony, 110
Clark, Bryan, 234
Clearwater, FL, 233
Clemson University, 31
Cleveland Browns, 100, 211
Cleveland Indians Progressive
   Field, 211-212
Coca-Cola®, 6, 36, 54-55, 77, 222-244
Coffel, Dennis, 71
College of Saint Rose, 98-99
College Swimming Coaches
   Association of American, Inc.
   (CSCAA), 101
Collison, Dave, 71-72
Colorado Rockies, 116
Colorado Springs, CO, 101
Columbia High School, 177-178

Columbus Crew Soccer Stadium, 210
Columbus, OH, 209-210
Comerica Park, 104
Concord, NC, 241
Condie, Richard, 117
Coors Baseball Field, 115
Copley, OH, 211-212
Cornell University, 171
Cosentino, Anthony P., 95-96
Cottage Grove High School, 129
Crochet, Kirk, 71

Dahl, Faye, 185
Dallas, TX, 181-182, 184, 194
Davino, Al, 139
Davis, Dennis, 195
Davis, Steve, 96-97
Davis, Steven, 129-130
De Avila, Lorena, 146-148
Deardoff, Carl, 249
DeCarolis, Bob, 99
Decker, Kurt, 216
Dedman, Robert, 109
Defries, Danny, 120-123
Degen, Becky Miser, 110
DeKalb, IL, 10-12, 19, 43, 240
Delano, Michael J., 104
Denver, CO, 67
Denzinger, Greg, 209-211
Detroit Red Wings, 204
Dicks Sporting Goods Stadium, 116
Dieck, Scott, 67, 102, 115-116, 243
Dimichele, Bob, 212
Dininger, Dave, 172
DiRocco, Michelle, 229
Disney Wide World of
   Sports®, 236-237
Dolphin Stadium, 234-235
Donini, Jesse, 139
Doolittle, Randy, 103-104
Doremus, Fred, 139

Douglas County Fairgrounds, 107
Downs, James M., 102
Dreibelbis, Mark, 104
Duluth, MN, 158
Durston, Mike, 176

Earth City, MO, 159, 162
East Rutherford, NJ, 215
Eastern Illinois University, 107-108
Eden Prairie, MN, 154-155, 157-158
Edgemont, PA, 229
Edmondson, Steven, 201
Ekeren, Dean M., 98
Elk River Ford, 157
Ellerbruch, Will, 101
Emerson, Rick, 170
Erickson, Jenny, 14, 99
Espinoza, Marcos, 140
Estrada G., Luis Carlos, 110
Even, Sarah, 250
Everett Public Facilities District, 108-109
Everett, WA, 108-109
Evers, Matt, 126

Fair-Play Scoreboards, 149
Farley, Paul, 77, 99, 217-220
Farmington High School, 160-161
Faurot Field, 162
Feasibility Study, 15-17
FedEx Field, 217
Feil, Chet, 25-41, 61, 63
Feil, Ruth, 25
Finish Lynx, 160
Firth, NE, 168
Fischer, Bob, 105
Fitzgerald, Dennis, 101-102
Fjeldos, Larry, 10
Flath, Brad, 236
Folmer, Brent, 170

Ford Center Arena, 87
Ford, John, 71, 219
Forunda, Ryan, 143
Fossell, Nate, 96
Foster, George P., 104
Fox, Michael, 102
Frechner, Bob, 139
Freshoul, Lynn, 71, 131
Fryn' Pan Family Restaurant, 105
Fulton Jr. High School, 200

Gadus, Ron, 71
Gainesville, FL, 103
Gallaudet, 76
Gallegos, Dan, 127-128
Garcia, Osiris, 191-193
Gasper, Jo, 123
Gator Bowl, 103
Gatzke, Brian, 15-17, 19, 34
Gatzke, Carla Kurtenbach, 43-46
Generations Restaurant, 101
George F. Emma Jr. Memorial Field, 15
George Long Company, 71
Gershman, Bob, 71
Giants Stadium, 215-216
Gibbons, Mike, 107, 139-142
Gida, Stan, 71
Giltspur, 99
Giorgio, Tom, 71, 231-232
Glanzer, Lisa Kurtenbach, 84, 180
Glendale Arena, 95
Godoy, Michael, 101
Gold Coast Hotel and Casino, 123
Goter, Neil, 111
Grace High School, 117
Grace, ID, 117
Gramm, Gary, 15, 17-18, 20, 48, 50, 57-59, 63-64, 68, 70-71, 75, 84, 91, 107, 118, 225
Grand Casino, 106

Grand Junction Park and
　　Recreation, 117
Grand Rapids, MI, 203, 205
Grant, Steve, 154
Grave, Perry, 61, 72, 82-83, 85-87,
　　180-181, 188
Green, Ken, 71, 159, 162
Gruner, Rocky, 138-140
Gullickson, Heidi, 203
Gullickson, Jeff, 69-70, 203-205

H.H. Dow High School, 107
Haas, David, 98
Hagele, Pat, 106-107
Haggerty, Mike, 225-226
Hale, Michael W., 118
Hank, Don, 141
Hansen, Don, 111
Hansen, Seth, 110
Harlingden Independent School
　　District, 191
Harman, Rob, 106
Harrell, Rob, 238-240
Harrison High School, 163
Hastings, NE, 167
Hattiesburg Dixie Youth
　　Baseball, 179
Hattiesburg, MS, 177-178
Hawke, Lyndi, 250
Hayward, CA, 142
Heine, Joel, 184-185
Heinse, Matt, 77
Heinz Field, 21
Hempfield High School, 230
Hendin, Nathan, 159-160
Hensley, Lori, 140
Hereid, Ginny, 101
Hermann High School, 160
Hess, Alan, 170-171
Hess, Jim, 71
Hickory Crawdads, 98

Hicks, Joseph (Joe), 163-164
High School Park and Recreation, 48,
　　113, 115 -117, 148, 167, 177
Holler, LeAnn, 108
Hollister High School, 163
Holsclaw, Bob, 71, 117 -119, 132
Hoosier Dome, 66-67, 102
Houser, Scott, 191, 204-207
Houston, Brad, 131-132
Houston, TX, 184-186, 193
Howell, Bob, 167, 169
Howell, Michael, 181-184
Howk, Brian, 219
Hubert H. Humphrey
　　Metrodome, 103
Huizenga, John, 71
Hulsebus, Joey, 154-155
Hurley, Chuck, 65

Indiana Athletic Directors, 64
Indiana University, 197-199
Indianapolis, IN, 61, 63-67, 102
Indianapolis Indians, 68, 108
Indianapolis Motor Speedway, 68
Invesco Field, 115
Iowa State Cyclones, 149

Jackpot Junction, 157
Jacobson, Bryan, 110
Jacobson, Scott, 102
James Madison University, 7
James, Jay, 98
Janesville, WI, 169
Jofer, Brent, 106
Johnson, Eric, 124-126
Johnston Public School, 149
Jones, Marlo, 28-43, 47-56, 60-61,
　　63-65, 68, 72, 74-75, 82, 91, 107,
　　110, 132-133, 150
Jonesboro, AR, 173

Jordan and McCallum Insurance, 97
Jordan, Matthew, 97
Junior College World Series, 117

Kansas City Royals, 152-153
Kansas University, 153
Karavedas, Nick, 119
Kelly, Marty, 170
Kempany, Mike, 236
Key Arena, 74-75, 133-134
Kinney, Steve, 71, 201-203
Kinney, Teresa, 201
Kleinjan, Barbra, 131, 148, 150
Knight, Bobby, 199
Knoxville, TN, 105
Koch, Ed, 143
Koenig, Kelly, 102, 158
Koerner, Jodi, 80
Koerner, Ron, 80
Kolar, Clay, 144
Kracji, Karen, 139-140
Krautter, Richard, 148
Kreeger, Roman, 147
Krogman, Jill, 222
Kurtenbach, Carla (see Gatzke)
Kurtenbach, Dr. Aelred, 3-25, 34-35, 51-52, 57-61, 66-69, 72, 74, 171, 180, 220, 225, 248
Kurtenbach, Frank, 13-15, 26, 33-34, 57, 66, 69-71, 90, 185
Kurtenbach, Irene, 26
Kurtenbach, Lisa (see Glanzer)
Kurtenbach, Paul, 73, 77, 79-81, 225-229

Lamar Advertising Company, 169-171
Lamar High School, 190
Lanphier High School, 146
Larson, Amanda, 131
Las Vegas, NV, 123

Lawler, Steve, 12
Lawson, Amy, 201
Learning World, 40
Lee, Jonathan, 109
Lehi, UT, 116, 130-132
Lenexa, KS, 150, 152, 154
Lenners, Rusty, 106
Lewandowski, Randy, 108
Lexington, KY, 200-201
Lexington, SC, 61, 67, 70-71, 241-244
Lighthouse Christian High School, 119
Lincoln, NE, 167
Linfield College, 96, 129
Little Creek Casino, 135-136
Little Rock, AR, 173
Long, George, 71-72
Los Angeles Memorial Coliseum and Sports Arena, 109
Loughran, Jim, 71
Louisiana State University, 177
Louisville, KY, 202
Lowell, Jeff, 137
Lower Columbia College, 96
Lowery, Maxine, 191
Lubbock, TX, 186, 191
Lundberg, Matt, 241
Lutsche, Sandi, 225

M Resort, 126
Mac Daddy's, 156
Madison, WI, 12
Mahopac High School, 217
Maki, Steven C., 103
Malibu, CA, 110
Mallanda, Al, 71
Mandalay Bay, 123
Mangus, Chris, 191
Marks Display System, 110-111
Marnell Corrao Associates, 126
Marsh, Dave, 61, 98, 163

Martin, Jack, 111
Masse, Telly, 157
Massillon High School, 213-214
Master's Academy, 238
Maxson, Cory, 118, 130-133
Maxwell, Steve, 195-196
Mayhew, Mike, 113-115
Mayo Civic Center, 96
McAdams, Clark, 215-216
McAlcester High School, 87
McCombs, Terry, 71
McDonald, Gary, 12
McDonald, Sean, 101
McDuffie, Richard, 107-108
McKean, Dick, 70-71
McShannock, Dan, 107
Meginnes, Mary, 144
Melissa Book Stadium, 206
Mercer Island Beach Club, 135
Mercer Island High School, 137
Mercer, WA, 137
Mesquite High School, 182
Mewis, Steve, 170
Miami, FL, 234
Michigan School for the Deaf, 208
Midland, MI, 107
Mihal, Daryl, 211-214
Mile High Stadium, 116
Miller, Sandra, 118
Minneapolis, MN, 103, 155-158
Minnesota Department of Parks and Recreation, 100
Minute Maid Park, 184
Mirage, 126
Misar, Becky, 110
Mizzou Arena, 162
Modesto, CA, 130, 142
Moen, Jason, 106
Montana Department of Transportation, 121-122
Montana Tech, 121
Montana, Joe, 199

Montgomery, Steve, 71, 190
Moore, Tim, 71, 162
Mooresville High School, 195-196
Morgan, Jim, 4, 9, 68, 100, 102-104, 106, 108, 118-119, 139, 247
Morris, Bryan, 174-176
Morry Stillwell, 110-111
Mountain Grove High School, 164
Mountain Grove, MO, 164
Mountain View High School, 174
Musa, Mario, 215
Muzzy, Nikole, 122

Naasz, David, 186
Nagel, Bryan, 61, 72-81, 111, 144, 159-160, 162, 225-229
Nagel, Teresa, 159
NASDAQ Info Center, 220-221
Nashville, TN, 245
National Collegiate Athletic Association (NCAA), 198
National Display Systems, 181
National Football League (NFL), 177
National Sign and Marketing Corporation, 138-140
Nationwide Arena, 210
Navasota Independent School District, 185
Nelson, John, 250
Nevco, 53-55
New Balance Track and Field Center, 102
New Orleans, LA, 177
New Palestine High School, 195
New York-New York Hotel and Casino™, 123
Newcomb, Bill, 71
Newcomer, Art, 69, 71, 154
Newman, Butch, 71
Nichols, Tim, 169-171
Niles, Carol, 204

Niles, MI, 203-205
Norman, OK, 84, 180
NorPac, 40, 54
Norris High School, 168
Novi, MI, 207

Odessa College, 186
Oklahoma State University, 87
Omaha, NE, 167
Oolman, Jen, 250
Orange Bowl, 234
Oregon State University, 99-100
Oriole Park, 108
Orlando, FL, 236
Osborne, Caitlin, 251
Ozark High School, 163-164
Ozark, MO, 164

Palazzo, 123-126
Paloma, John, 67, 69, 194, 196, 198-200
Papaco, Rudy, 140
Paris, Jeff, 71
Parker, Chris, 170
Parma City School District, 213-214
Parsons, Barry, 71
Parsons, Kathy, 139
Pattonville High School, 162
Paul, ID, 132
Payton, Walter, 178
Penn State, 77
Pepsi-Cola®, 54-55, 77, 120, 132, 210, 244
Perry, Staci, 251
Pete Ragus Aquatic Center, 190
Pheonix, AZ, 106, 113
Philadelphia, PA, 229
Phillips Arena, 240
Phillips, Harold, 107
Phoenix Municipal Stadium, 106

Phoenix Sky Harbor Airport, 114-115
Pistons, 204
Plainsfield High School, 198
Plant, Michael, 106
Plum, Bobby, 198
Pocahontas Area Athletic Booster Club, 101-102
Pocahontas, IA, 101
Poe, Don, 71
Poppell, Ed, 103
Portland, OR, 129
Preston, Jim, 136, 150, 152, 154
Pufall, Jerry, 109
Purdue University, 65
Puthoff, Meghan, 201
Puthoff, Ted, 201

Ramirez, David, 220-224
Ramsbacher, Randy, 127
Raymond James Stadium, 233
Reed, Hub, 71
Reeves, Kelly, 71
Rendon, Charles, 140
Republic, MO, 163-164
Richer, Joe, 71, 110, 137
Ripken Stadium, 226-227
Robbins, Gudrey, 12
Robbins, Jeff, 12, 19, 22
Robert F. Kennedy (RFK) Memorial Stadium, 217
Robertson, Gene, 71
Robles, Israel, 139-140
Rochester, MN, 158
Rocky Hill, CT, 215-216
Rodriguez, Jose, 191
Roland, Kirc, 97
Roseburg, OR, 107
Ross, Gary, 149
Rowe, Bill, Jr., 98
Roy, Dan, 233
Rozie, Dany, 141

SAFECO Field, 133-134
Safstrom, Fred, 108-109
Saginaw County Event Center, 101
Saginaw, MI, 101
Salt Lake Community College, 103
Saluki Swim Club, Inc., 144
Samsung®, 224
San Antonio, TX, 61, 187-189, 191
San Diego Padres, 108
San Francisco, CA, 130
Sander, Dr. Duane, 3-6, 18, 65, 248
Sander, Norbert W., Jr., 102
Santa Clara Sports Complex-Aquatic Center, 106-107
Sarmiento, Victor, 235
Saudi Arabia, 31, 33
Schumacher, Dick, 71, 154
Schwan, Pat, 10
Scoreboard Parts and Service, 26-29, 34-35
Scoreboard Sales and Service, 17, 37, 41-50
Scottrade® Center, 161-162
SeaTac Airport, 36
Seattle Mariners Kingdome, 36, 38
Seattle Pacific University, 39
Seattle Supersonics, 136
Seattle, WA, 2, 25-55, 58, 60-61, 63-65, 68, 72, 74, 82, 91, 107, 110, 133-134, 136, 150, 225
Selberg, Gregg, 106
Sheridan, CO, 115
Sherryland Independent School District, 191-192
Shrewsberry, Kristopher, 212-213
Shull, Melba, 243-244
Siebert, Randy, 195
Siemonsma, Tom, 154-155, 158
Simon, Kurt, 111
Sioux Falls, SD, 105
Skagit Valley Casino Resort, 103
Skare, Adam, 154-158

Skutt Catholic High School, 167
Smiley, Steve, 71
Smith, Lynette, 47, 57, 91
Snook, Jason, 207-209
Snoqualmie Pass, 105
Sosa, Kevin, 140
South Dakota State University, 150
South Mountain Community College, 114
Southern Illinois University, 144
Southwest Missouri State University, 98
Southwest Texas University, 92
Spectrum, 185, 187-189
Spencer County High School, 203
Spetz, Tony, 170-171
Spokane Coliseum, 36
Spokane Valley, WA, 110
Spokane, WA, 11
Spotlight Live, 221
Spring Valley City Bank, 15
Springfield, IL, 146
St. Bonifacius, MN, 157
St. Cloud, MN, 156
St. Louis Blues, 161-162
St. Louis University, 161-162
St. Louis, MO, 76-77, 81, 102, 159, 162,
St. Paul, MN, 156
Starrsmill High School, 241
State Capitol Building, 34
Steele, Jeffrey, 96
Steinberg, Charles A., 108
Steinkamp, Mark, 50
Steinlicht, Kelly, 98
Stephens, Harold, 71
Stoll, Valorie, 173
Stopple, Mark, 170
Strafford R-VI Public Schools, 166
Strand, Al, 71
Strasburg, ND, 12
Sugar Bowl, 177

Super Bowl, 199
Sutherland, Roy, 96
Swiftel Center, 172
Sydow, Kyle, 148

Tacoma Raniers, 56
Tacoma Tigers Baseball Club, 104
Tacoma, WA, 104
Taylor, Bruce, 71, 97
TDK, 220
Texas A&M, 98
Texas Christian University, 190
Texas House of Representatives, 92
Texas Longhorns, 5, 190
Texas Memorial Stadium, 5
Texas Rangers, 183
Texas Tech University, 186-187
The Racquet Club of Memphis, 246
Thiner, Dany, 61, 63-72, 148-150, 194, 196-197
Thorness, Greg, 11-12
Thurman, Michael, 103
Times Square, 5, 6, 220-223
Topeka Sports Park, 153
Topeka, KS, 152-153
Toyota Center, 184
Treasure Island, 126
Tressel, Jim, 210
Trevino, Louisa, 140
Tucker, Brian, 106
Tulsa Union High School, 87
Turfway Park, 201
Turner, Charles, 70-71
Twin Falls, ID, 119

U.S. Naval Academy, 77, 79
U.S. Olympic Festival, 102
Unger, Mike, 195
Union School District, 165
Union Stewards, 12

United Dairyman, 119
United States Figure Skating Association (USFSA), 110-111
University High School, 110
University of Arkansas Razorbacks, 109
University of Denver, 110
University of Florida, 103, 233-234
University of Georgia, 233, 239
University of Illinois, 145-146
University of Iowa, 149
University of Kansas Highland Ballpark, 153
University of Kentucky, 68
University of Louisville, 68
University of Maryland, 77
University of Miami, 234
University of Michigan Hockey Arena, 204
University of Missouri, 162
University of Montana, 121
University of Nebraska, 31
University of New Mexico, 127-128
University of Northern Iowa, 149
University of Notre Dame, 199, 204-206
University of Oklahoma, 85, 87, 180-181
University of Southern Mississippi, 177
University of Tennessee, 105
University of Texas, 5, 190
Utah Legislature's Electronic Voting System, 2-3

Van Cleff, Rusty, 139
Van Ess, Janet, 124-125
VanBemmel, Al, 41
VanLeeuwen, Diane, 206
VanSickle, Ken, 110
Vargas, Jose, 144-146

Vasgaard, Jim, 19, 173
Vaught, Ron, 139
Vela, Pete, 71, 191-193
Vella, Allen C., 101
Venetian® Hotel and Casino, 123-125
Victory Field, 108
Vogel, Jason, 201
Vogelgesang, Dave, 71-72
Vugteveen, Paul, 102

Waco ISD Sports Complex, 182
Wagner Community School District, 111
Wagner, Justin, 217-220
Wagner, Scott, 218-219
Wagner, SD, 111
Wagoner, Chris, 130, 142-143
Waldmann, Mike, 190
Washington Department of Transportation, 105
Washington House voting console, 2, 4
Washington Redskins, 77-78
Wasserman, Ed, 140
Webb, Jim, 71
Weinacht, Keri, 106
Weldner, Steve, 123
Wendler, Brett, 188
Weninger, Deanne, 10-12

Weninger, Ed, 9-12, 19, 30-31, 34, 43, 47, 50
Weninger, Lanette, 12
Weslaco Independent School District, 191-192
Weslaco, TX, 191-193
West Minico Middle School, 118, 132
Westminster Presbyterian, 157
Westphal & Company, 170
Wheelersburg Cinema, 210
Wheeling, IL, 146
Whittington, Scott, 200-202
Wichita State University, 153
Wildeman, Paul, 34, 61, 72, 89-92, 187-191
Williams, Kyle, 60, 133-137
Wilson, Greg, 217, 231
Woodmont High School, 237
Wright State University, 68
Wyant, Gary, 105
Wynn Las Vegas, 126

Xavier University, 99

Yarmark, Brad, 171
Yoo, Jee, 140
Yost Arena, 204